Cambridge Studies in Biological Anthropology

Series Editors

G. W. Lasker
Department of Anatomy, Wayne State University, Detroit, Michigan, USA

C. G. N. Mascie-Taylor
Department of Biological Anthropology, University of Cambridge

D. F. Roberts
Department of Human Genetics, University of Newcastle-upon-Tyne

Cambridge Studies in Biological Anthropology

Biological aspects of human migration

Biological aspects of human migration

EDITED BY

C. G. N. MASCIE-TAYLOR

Lecturer in Biological Anthropology and Fellow of Churchill College, Cambridge
Department of Biological Anthropology
University of Cambridge

AND

G. W. LASKER

Emeritus Professor of Anatomy and Adjunct Professor of Anthropology
Department of Anatomy
Wayne State University
Detroit, Michigan

The right of the
University of Cambridge
to print and sell
all manner of books
was granted by
Henry VIII in 1534.
The University has printed
and published continuously
since 1584.

CAMBRIDGE UNIVERSITY PRESS

Cambridge   *New York*   *Port Chester*
Melbourne   *Sydney*

CAMBRIDGE UNIVERSITY PRESS
Cambridge, New York, Melbourne, Madrid, Cape Town, Singapore, São Paulo, Delhi

Cambridge University Press
The Edinburgh Building, Cambridge CB2 8RU, UK

Published in the United States of America by Cambridge University Press, New York

www.cambridge.org
Information on this title: www.cambridge.org/9780521118491

First published 1988
Reprinted 1989
This digitally printed version 2009

A catalogue record for this publication is available from the British Library

Library of Congress Cataloguing in Publication data

Biological aspects of human migration.
(Cambridge studies in biological anthropology; 2).
Includes index.
1. Man — Migrations   2. Emigration and immigration.
3. Human population genetics.   I. Mascie-Taylor, C. G. N.
II. Lasker, Gabriel Ward.   III. Series.
GN370.B56   1987   304.8   87-8055

ISBN 978-0-521-33109-8 hardback
ISBN 978-0-521-11849-1 paperback

Contents

List of contributors   vii

1 The framework of migration studies   1
   G. W. LASKER AND C. G. N. MASCIE-TAYLOR

2 Peopling of the continents: Australia and America   14
   W. S. LAUGHLIN AND A. B. HARPER

3 Migration in the recent past: societies with records   41
   D. F. ROBERTS

4 Models of human migration: an inter-island example   70
   P. D. RASPE

5 Rural-to-urban migration   90
   BARRY BOGIN

6 In search of times past: gene flow and invasion in the   130
   generation of human diversity
   K. M. WEISS

7 Migration and adaptation   167
   MICHAEL A. LITTLE AND PAUL T. BAKER

8 Migration and disease   216
   BERNICE A. KAPLAN

   Glossary   246

   Index   248

Contributors

Professor P. T. Baker
Department of Anthropology
The Pennsylvania State University, University Park, Pennsylvania
16802, USA

Dr B. Bogin
Department of Behavioral Sciences
University of Michigan, Dearborn, Michigan 48128, USA

Dr A. B. Harper
Department of Biobehavioral Science
University of Connecticut, Storrs, Connecticut 06268, USA

Professor B. A. Kaplan
Department of Anthropology
Wayne State University, Detroit, Michigan 48202, USA

Professor G. W. Lasker
Department of Anatomy
Wayne State University, School of Medicine, 540 Canfield Street,
Detroit, Michigan 48201, USA

Professor W. S. Laughlin
Department of Ecology and Evolutionary Biology
(Box U-154), University of Connecticut, Storrs, Connecticut
06268, USA

Professor M. A. Little
Department of Anthropology
State University of New York, Binghamton, New York 13901, USA

Dr C. G. N. Mascie-Taylor
Department of Biological Anthropology
University of Cambridge, Downing Street, Cambridge CB2 3DZ, UK

Dr P. D. Raspe
c/o Department of Biological Anthropology
University of Cambridge, Downing Street, Cambridge CB2 3DZ, UK

Professor D. F. Roberts
Department of Human Genetics
University of Newcastle-upon-Tyne, 19 Claremont Place,
Newcastle-upon-Tyne, UK

Professor K. M. Weiss
Department of Anthropology
Pennsylvania State University, 409 Carpenter Building,
University Park, Pennsylvania 16802, USA

1 The framework of migration studies

G. W. LASKER AND C. G. N. MASCIE-TAYLOR

Introduction

The impact of migration on human biology is of considerable importance to a wide variety of disciplines including anthropology, demography, epidemiology and genetics. Migration simply means movement of individuals or groups of individuals. To say that someone is a migrant implies emigration from one population and immigration into another. Migration usually refers to spatial (geographic) movement although it is also used to define social or occupational mobility. The authors of this book intend the narrower definition and 'migration' here refers to geographical displacement of people: changes of residence involving introduction of individuals into a different locale. If the extent, direction and orientation of the movement are not specified, migration remains qualitative and the differentiation is merely between migrants and non-migrants.

The biological aspects of migration have more to do with the consequences than with the causes of such movements of people. Therefore the focus of this book is not so much on why people choose or are forced to move but on the biological implications of that movement to both recipient and donor populations. Consequently, although labour migration is important socially and economically, it is only considered here (see Chapter 8) if there is some adaptive or maladaptive component to movement. Likewise the movement of refugees is considered only from the point of view of the results of biological stresses that cause or are incidental to the displacements and we shall not deal with the social aspects of human suffering caused by refugeeism, this common outcome of forced mass migration.

The magnitude of migration can vary and it is common to distinguish international, national (usually referred to as internal), local, rural and urban migrants. In general, international migration implies a greater spatial movement but this need not always be true.

Independent of the migrational level is the type of migration. Biological anthropologists and geneticists talk about random and selective migration. This distinction is important when considering the biology of the donor population and interpreting differences in biological conditions between them and the migrants some time after the event by studying the conditions of the sedentes (non-migrants) who stayed behind. From a genetical point of view, migration may result in changes in gene frequencies in the population concerned if migration is also associated with gene flow. Gene flow relates to the exchange or influx of different genes caused by migration by individuals of a species which has different gene frequencies in different places. Changes can result from very short-distance migrations provided that they involve a large proportion of the population and are repeated generation after generation. Consequently, migration and gene flow are important evolutionary factors. Random migration of genes will lead to or promote genetic homogeneity between populations or subpopulations whereas selective migration can promote or maintain differences between populations.

The questions approached through human migration studies

In human biology there are two virtually independent lines of migration study.

(1) Human migration is the mechanism that injects DNA from one gene pool into another. Premarital migration, marital migration, and postmarital migration can each have the effect of altering the composition of the recipient population's genetic endowment. The effects of selective emigration have not received much attention because selective emigration in respect to hereditary traits is not known to be appreciable, except in very small populations where the results would be similar to those of random genetic drift. Problems with studies of migration as a transporter of genes involve such issues as the lack of clear boundaries of human populations, and difficulties in determining the genotypes of migrant and non-migrant individuals in respect to characteristics of interest.

(2) Human migration is also the mechanism that inserts similar kinds of individuals into diverse environments. Thus, for the study of environmental influences on human beings, the comparison of migrants with non-migrants of identical (or at least similar) genotypes provides a statistically sound basis from which to analyse the influence of the factors that differ between the environments of the places which donate emigrants and those which receive immigrants. Problems with this kind of study, as pointed out by Kaplan (1954), lie in determining which of a complex array of environmental variables that differ between donor and receiving places

are responsible for effects found in the biological comparison of migrants with non-migrants.

(3) Theoretically, there should be a third approach: studying the interaction of genetic and environmental results of migration. In fact, however, in order to study the genetic effects of migration it has been necessary largely to confine attention to traits of high heritability such as blood groups, serum proteins and the like, whereas in the study of environmental effects a different set of variables such as anthropometric and physiological measurements has been used. The chief area in which the interaction of the two influences may be sought is probably in respect to human diseases of mixed aetiology. The study of sickle-cell carriers among descendants of migrants from malarial to non-malarial environments exemplifies one class of simple interaction study. The difficulties of extending even this model to studies of thalassemias, G6PD deficiency and Duffy blood groups show how complex will be the study of other conditions of mixed aetiology.

History of migration studies

The study of how migration has influenced regional differentiation and convergence of human populations in respect to biological distinctions is as old as an identifiable science of physical anthropology. Unfortunately the conceptualization of the role of migration on human biology has often been crude and sometimes downright misleading. At the time when almost every characteristic of the skeleton was considered to be an inherent racial hallmark, the geographical distribution of such characteristics was assumed to result solely from movement of peoples (races) who possessed the characteristics. When these characteristics appeared in modified form or reduced frequencies, the modifications or less marked occurrences of the traits were arbitrarily ascribed to race mixture. There were some hints of evidence that did not fit the pattern. Fishberg (1905) reported that migrants had offspring different in stature from themselves; apparently the children had grown up to be different from their parents because of a biological response to the different environment. Franz Boas presumably knew of these studies and he set out to test, on adequately sized samples and with the best available controls and statistical methods of the time, the thesis that the differences resulted from improvement in conditions rather than from some unspecified selection of genetically determined traits. Boas' (1910, 1912) own extensive studies of the results of migration to the United States of Old World Jews and Sicilians is, of course, a classic. It demonstrated, as did the studies by others that confirm

its findings (e.g. Shapiro, 1939; Ito, 1942; Goldstein, 1943), that stature
and other anthropometric measurements are modified in the offspring of
migrants. Basler (1927) showed that it made a difference to the shape of
the head, as measured by the cephalic index, whether an infant had been
swaddled and placed on its back or whether it was placed on its side and
Ewing (1950) found that migrants who abandoned the practice of swaddl-
ing had children who became much narrower headed than themselves.
The role of migration in human biology through its moving of DNA from
one place to another (in the gametes of migrant individuals) and through
subjecting the products of DNA to environments that were effectively
different (in the sense that attributes of human organisms are products of
DNA) is subject to qualitative differences in the human migrations.
Migration can be an individual matter, a family matter or a group matter.
These and their subtypes have varying biological implications. An individual migrant may transport cultural ideas which have significant biological implications (food preferences, for instance) but he or she is likely to
be absorbed into the population of the receiving place with little influence
on its gene pool and even less on the population left behind. Numbers of
people, migrating together, can have quite different aspects. Groups can
occupy empty territory and transplant a more or less representative
culture and a sample of their DNA into a new environment where
different physical conditions prevail. Alternatively, migrating groups
may swamp the receiving population and/or deplete the donor one in
respect to selected genetic or other traits.

Given such a variety of migration types, it is hardly surprising that
anthropologists put quite diverse emphases on migration, conduct their
studies in different settings, and reach 'conclusions' that are diverse. Only
a combination of such studies and a synthesis of the points of view are
likely to give a rounded image in which the biological outcome can be seen
to depend on the differences between donor and recipient people,
cultures and places, the numbers of migrants and their ages and sexes.

A contrast exists between a view of migration as an historical event and
a view of migration as steady pressure constantly making for greater
genetic similarity among populations – a process that is kept in equilibrium short of panmixia by other evolutionary forces such as differential
selection pressures in different parts of the human ecosphere. The first
view is generally based on the assumption of strong genetic determinism.
If the peoples of the world (proponents of the view would generally say
'races', not 'peoples') remain biologically the same for indefinitely long
periods, virtually independently of cultural and physical environments,
then biological similarities could be ascribed, it has been argued, to

common origins. When people with some similar traits are observed at some remove from each other, proponents of this view invoke hypothetical migrations to have gotten them there. Typically, one of a pair of racial populations is thought of as newer than the other and hence the new one must be a migrant one. This kind of thinking was prominent in most of the older accounts of the peoples of the world. For instance Hooton (1930) identified among the Pueblo Indians types that reminded him of various peoples from other parts of the world, and Birdsell (1949) hypothesized that there were three distinct migratory streams into Australia of Carpentarian, Murrayan and Cairnsian types. Some modern workers continue to speculate about specific places and times of hypothetical migrations, but usually in more restrained terms that take into account a fuller record of the geography, geology, palaeontology and archaeology of the region. Such a 'historical' approach to human migration varies in value according to the amount and quality of the documentation. Because of the usual dearth of evidence surviving from the time of the supposed migrations, much of the interpretation is founded on the comparison of biological characteristics of now-living individuals. It is difficult to synthesize from these accounts general principles about human similarities and differences; the origins of these tend to be pushed back to hypothetical events in a little-known and distant past.

The pressure of repeated migration

An alternative approach to migration studies may be characterized as generalizing rather than historical. It seeks to discover regularities in human migratory behaviour and its biological results. Differentiation and assimilation are considered, in this view, to continue over time. Proponents of the view draw inferences from other biological species, and build models and test alternative ones, applying them to the sparse data on biological effects of human migration.

Again the concern has been largely with inherited traits or the heritable component of other traits, but the models are evolutionary and imply that environments have a role through natural selection on the migrants. In this approach migration is classified according to origin, direction and distance. By assuming that origins are point origins, directions are equiprobable and distances follow some law of physics, the law can be applied to the biological phenomena. The law of brownian movement can be applied to dispersal of organisms as well as the diffusion of particles, and the law of gravitation can be applied to intermating between populations as well as the attraction to each other of separated masses. Simple models can be constructed to test the degree of conformance of empirical data on human migration to such theoretical principles. The extent and nature of the deviations from such a model permit the addition of subsidiary modifications to the model and for further, but well-circumscribed, speculation under the historical rubric about the influence of unique past events. The impetus to this model-building came through population genetics (Malécot, 1948) and has been applied to human populations by Morton (1977) and others. Wijsman & Cavalli-Sforza (1984) have recently reviewed models of individual migration, isolation by distance, gene flow, clines and demic diffusion and assess the state and prospects for theories of human migration that expose in mathematical terms the results of having moved genes about in geographical space. Empirical exploitation remains a fertile field for future research. Long-distance and short-distance migrations may be somewhat independent of each other in frequency of occurrence and direction, perhaps because of different modes of transportation for journeys of different distances. These and other issues (such as the role of geographical features in channelling human movements) can be tested in specific situations. Empirical studies should help to refine knowledge of the primary and secondary conditions that account for the genetic effects of human migration. For example, premarital, marital and postmarital migration have different implications since a couple migrating to mate may thereby lead to recombination of genes sampled from two different gene pools, whereas a couple migrating together postmaritally will have offspring comparable to those of couples left behind, but in a different place.

Prospective study of the effect of migration on gene flow

An example of the effect of migration on the gene pool comes from some recent work by Mascie-Taylor & Lasker (1987) and Mascie-Taylor & Boyce (in press). They examined the distribution of the ABO and Rh blood groups in Britain before and after migration. Since they were dealing with the same people at different times, any changes in the blood group distributions must have been caused by migration. It is well known that ABO and Rh blood groups show clinal variation with a tendency for O-gene frequency to decline from Scotland to southern England as well as from Wales to East Anglia. Rh-gene frequency decreases from East Anglia to Wales (Kopeć et al., 1970).

In these recent studies, the sample comprised the mothers of children who were members of the National Child Development Study. This study commenced in 1958 when all children born in the week 3–9 March were examined. The children and their families were periodically restudied in 1965, 1969, 1974 and 1982. In 1958 the mothers' blood was typed for ABO (no subtyping of A) and Rh (+ or −) and the geographical location of the family recorded. The relationship between blood group distribution and geographical location was examined at three points in time: at the mothers' births (which averaged about 1930), at the time of birth of their child in 1958, and in 1974. The analyses were restricted to the 8850 mothers whose child was domiciled with them in 1958 and 1974.

It should be noted that the analyses only detailed the effects of internal migration. Even so the results tend to underestimate the true effect of this migration since migration within each geographical region studied is ignored and the reduction in the sample at each restudy is largely caused by families moving from their previously known address. Although all reasonable attempts were made to follow up families between sweeps this can never be 100% foolproof. In 1974 the response rate was still in excess of 88%.

The easiest way to determine the effects of migration is to calculate the number of generations it takes to reach a certain degree of geographical homogeneity (see Chapter 4). In these studies families were placed into one of the 11 standard regions comprising Scotland, Wales, the South-west, Southern, Midlands, London and South-east, East, North Midlands, North-west, North, and East and West Ridings of Yorkshire. Consequently the number of generations it took to reach 99% homogeneity between these 11 regions was determined. Although slight differences arise depending on which clustering algorithm was used, the results all indicated that for ABO blood group gene frequencies homogeneity of part of England with Wales is reached in about six 'generations', i.e. about 160 years based on the comparisons between mother's birth and 1974 (see Fig. 1.1). When the analyses were restricted to the period birth to 1958 then that degree of homogeneity is reached in about nine 'generations', i.e. 250 years, where a generation (from birth of mother to birth of child) is approximately 28 years. The time taken for homogeneity to occur throughout Britain is considerably longer. This is because there are major gene frequency differences between Scotland and the rest of Britain. Under one set of assumptions, 99% geographical homogeneity takes approximately 500 years to attain.

Also examined was the geographic distribution of the blood group phenotypes according to the place of residence of the mothers in 1958 and in 1974. For this purpose the centre of population of the county was used (or, in the cases of Scotland and Wales, group of counties) as an approximation of place of residence, so only intercounty migration was

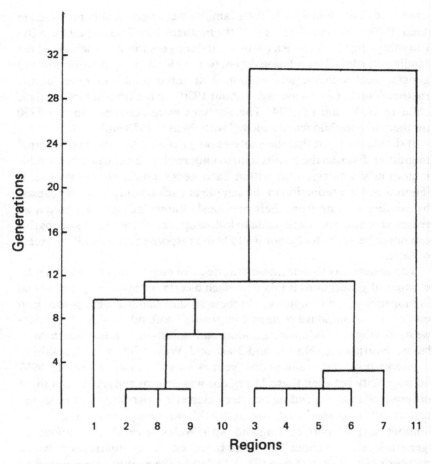

Fig. 1.1. The length of time in generations it would take for any two regions of
Britain to become 99% homogeneous genetically if the rate of migration between
regions observed in the cohort under study was continued. 1 = North West,
2 = North, 3 = East and West Ridings of Yorkshire, 4 = North Midlands,
5 = East, 6 = London and the South East, 7 = South, 8 = South West,
9 = Midlands, 10 = Wales, 11 = Scotland.

recorded. Blood groups O, A, B plus AB and Rh− were separately
plotted against latitude and longitude by polynomial regression. When
both coordinates are considered together, each of these blood group
forms (AB being lumped with B because of the small numbers) showed a
statistically significant cline. It can be seen from the plots of examples of
these (Fig. 1.2) that the clines flatten with time as migration renders the
distribution of this cohort of 8850 fertile women appreciably more
geographically homogeneous in respect to blood group phenotypes.

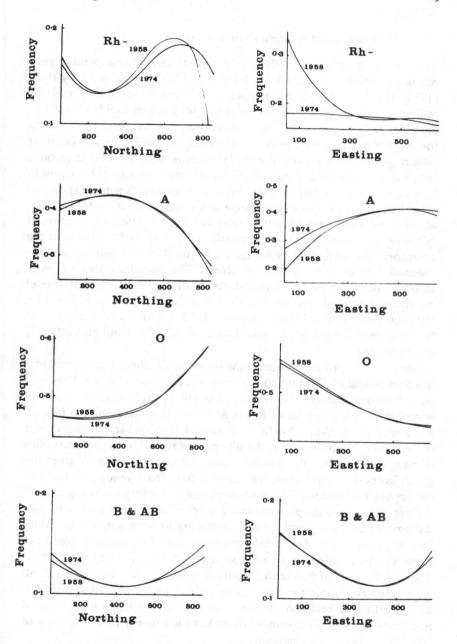

Fig. 1.2. Clines of blood groups in a cohort of British women to show the effects of intercounty migration during a 16-year period. Drawn from polynomial regressions of phenotype frequencies on northing and easting. Scaled in kilometres north and east, respectively, of the ordnance reference point.

Migration and changed environment

This development of the study of migration from a genetic point of view has been virtually independent of the line of enquiry initiated by Boas (1910, 1912) on the direct impact of the environments on migrant individuals. A model for such studies was set forth by Goldstein (1943). He extended Boas' comparison of those born in a receiving country with the immigrants to a four-fold comparison of (1) sedentes and (2) their offspring in the donor country, and (3) migrants and (4) their foreign-born offspring in the receiving country. Goldstein used a series of Mexicans in Mexico to serve as controls; their Mexico-resident offspring to show the effects of temporal (secular) change from one generation to another; Mexican migrants to the United States to show the effect of selection of migrants; and their American-born offspring to show the effects of the environmental impact of being raised in the different setting plus the generation-to-generation secular effect. The variables studied were anthropometric measurements and observations. The Goldstein model was later extended also to include the comparison of sedentes with returned emigrants in Mexico (Lasker, 1952, 1954). Other variations of the Boas model have been applied by Shapiro (1939) and Hulse (1957) among others.

The possibility of differences in the influences of similar environmental shifts on people of different constitution was implicit in Boas' selection of more than one immigrant group for his study. It has been more explicitly approached in the studies of Baker & Little (1976) of Peruvian Indians and Mestizos and White North Americans in highland and lowland areas of Peru. In general, however, the diversity of results of migration studies of many kinds of people in many places seems to result more from the great diversity of environmental impacts than from genetic diversity of the populations studied. The environmental variability has two aspects.

First, there is no simple dichotomy in geographical environments, nor are environments arrayed linearly according to some single dimension. Instead, temperature (at various seasons), humidity, altitude, wealth of resources (according to various systems of exploitation), vectors of diseases, sources of foodstuffs, hours of sunshine, inches of rainfall, all vary with geographic location. The combination of any migrant donor locale with any recipient locale plus the great variation in cultural responses to natural environments makes for a nearly infinite variety of types of migratory experiences.

The second reason for variation in the results of studies of the effects of human migration on similar aspects of biological response, is the wide variety of variables that have been tested. The original problem set by Boas was to submit the assumption of stability in racial characteristics to test. The chief such characteristic in the thinking of anthropologists in the 50 years from the time of Retzius' publication to the time of Boas', was the cephalic index. This and other anthropometric measurements of the living that correspond with craniometric and other osteometric dimensions were among the first traits studied in migratory situations by Boas' successors. Stature and other length measurements of body segments responded similarly to each other in migrants exposed to improved diet and general environmental conditions. This consistency led to the idea of a uniform 'environmental growth factor'. This simple notion broke down as the studies became more diverse.

Following Boyd's (1950) application of genetics to raciology, and especially after Mourant's (1954) publication of a world-wide survey of blood groups, it became clear that evolutionary changes in gene frequencies would ordinarily be very slow (the issue of neutrality to selection is still being debated) and that geographical distribution of genes would bear a relationship to preceding migration. Other migration studies turned to more labile traits in which one-generation changes (without genetic differentiation) could be expected to be found. These traits included such physiological variables as basal metabolic rate, vital capacities, blood pressure and ability to perform work, and other indices of pulmonary, cardiac, muscular and digestive responses to stresses of changed environments. Of the various results, the favourable response to life-long residence at high altitude in respect to pulmonary capacities is one of the most adequately documented findings. Development of special capacity to respond to extremes of temperature, demands of work, shortages of food, etc., remains somewhat problematical, although plausible in theory and supported by the results of some migration studies.

Prospects

How then can one summarize our conception of migration study in respect to biological traits? One can no longer speak of a single repeatable pattern. Instead, purely genetic and environmentally tractable traits must be distinguished (either before or by the studies). The purely genetic can be examined in respect to their transport by migration. The geographical distributions, at least among nearby places, is ordinarily much more the direct result of migrations than of any selection by the varying environments. The more tractable traits can be studied as to the kind and extent of modification of similar genotypes between various kinds of pairs of environments and with migration between them at various ages in the development cycle. Inconsistency of results must be examined in the light of adequacy of sample sizes, comparability of variables and the techniques by which they are measured, and similarity of the environments. Genetic differences among human populations in their capacities to respond to environmental changes are a less plausible explanation of most instances of diversity of results.

By now the chief question raised by the earlier studies has been answered and the diversity of results found by Kaplan (1954) has been reasonably explained. It is thus clear that migration brings into play forces adequate to modify phenotypes appreciably, but the diversity of environments leads to diverse results – not the single migration pattern postulated earlier. This has freed modern research to look at other kinds of variables in migrants such as excretion of catecholamines (James *et al.*, 1985), diabetes mellitus (Gerber, 1984) and obesity (Ramirez & Mueller, 1980). Even where the direction and extent of effects are reasonably well known, the mechanisms need to be revealed through the most characteristic of anthropological methods – comparison of studies conducted on peoples of different cultures and in different natural settings in respect to a wide (essentially open-ended) array of variables.

References

Baker, P. T. & Little, M. A. (1976). *Man in the Andes: a Multidisciplinary Study of High Altitude Quechua*. Stroudsburg, Penn: Dowden, Hutchinson and Ross.

Basler, A. (1927). Ueber die Einflusse der Lagerung von Sauglingen auf die Bleibende Schadelform. *Zeitschrift für Morphologie und Anthropologie*, **26**, 247–55.

Birdsell, J. B. (1949). The racial origin of the extinct Tasmanians. *Records of the Queen Victoria Museum, Launceston, Tasmania*, **2(3)**. Reprinted in: *Yearbook of Physical Anthropology* (1950), **6**, 143–60.

Boas, F. (1910). *Changes in Bodily Form of Descendants of Immigrants*. Senate Document 208, 61st Congress, 2nd Session, Washington, DC.

Boas, F. (1912) Changes in bodily form of descendants of immigrants. *American Anthropologist*, **14**, 530–62.

Boyd, W. C. (1950). *Genetics and the Races of Man*. Boston: Little, Brown.

Ewing, J. F. (1950). Hyperbrachycephaly as influenced by cultural conditioning. *Papers of the Peabody Museum of American Archaeology and Ethnology, Harvard University*, **23(2)**.

Fishberg, M. (1905). Materials for the physical anthropology of the Eastern European Jew. *Annals of the New York Academy of Science*, **16**, 155–297.

Gerber, L. M. (1984). Diabetes mortality among Chinese migrants to New York City. *Human Biology*, **56**, 449–58.

Goldstein, M. S. (1943). *Demographic and Bodily Changes in Descendants of Mexican Immigrants*. Austin: Institute of Latin American Studies, University of Texas.

Hooton, E. A. (1930). *The Indians of Pecos Pueblo: a study of their Skeletal Remains*. New Haven: Yale University Press.

Hulse, F. S. (1957). Exogamie et hétérosis. *Archives Suisse d'Anthropologie Général*, 22, 103–27. Translated into English in *Yearbook of Physical Anthropology*, 9, 240–57.

Ito, P. K. (1942). Comparative biometrical study of physique of Japanese women born and reared under different environments. *Human Biology*, 14, 279–351.

James, G. D., Jenner, D. A., Harrison, G. A. & Baker, P. T. (1985). Differences in catecholamine excretion rates, blood pressure and lifestyle among young Western Samoan men. *Human Biology*, 57, 635–47.

Kaplan, B. A. (1954). Environment and human plasticity. *American Anthropologist*, 56, 780–800.

Kopeć, A. C. (1970). *The Distribution of the Blood Groups in the United Kingdom*. Oxford: Oxford University Press.

Lasker, G. W. (1952). Environmental growth factors and selective migration. *Human Biology*, 24, 262–89.

Lasker, G. W. (1954). The question of physical selection of Mexican migrants to the USA. *Human Biology*, 26, 52–8.

Malécot, G. (1948). *Les Mathématiques de l'Hérédité*. Paris: Masson.

Mascie-Taylor, C. G. N., & Boyce, A. J. (1987). ABO Variation in Britain. *Annals of Human Genetics* (in press).

Mascie-Taylor, C. G. N., & Lasker, G. W. (1987). Migration and changes in ABO and Rh blood group clines in Britain. *Human Biology*, 59, 337–44.

Morton, N. E. (1977). Isolation by distance in human populations. *Annals of Human Genetics*, 40, 361–5.

Mourant, A. E. (1954). *The Distribution of the Human Blood Groups*. Oxford: Blackwell.

Ramirez, M. E. & Mueller, W. H. (1980). The development of obesity and fat patterning in Tokelau children. *Human Biology*, 52, 675–87.

Shapiro, H. L. (1939). *Migration and Environment*, Oxford: Oxford University Press.

Wijsman, E. M. & Cavalli-Sforza, L. L. (1984). Migration and genetic population structure with special reference to humans. *Annual Reviews of Ecological Systems*, 15, 279–301.

2 Peopling of the continents: Australia and America

W. S. LAUGHLIN AND A. B. HARPER

A comparison of the peopling of Australia and America offers insights into the population history of our species that would otherwise remain submerged. We begin with the observation that the first Americans have been in America some 15 000 years and they still resemble the Mongoloids of north-eastern Asia. In marked contrast, the first Australians have been in Australia well over 30 000 years and they do not resemble any of the mainland or Indonesian populations of south-eastern Asia.

The scope of essential topics and relevant data necessary to the comparison of the populations of these geographically disparate areas is now so large that it is necessary to organize the main topics or themes, and to condense and synthesize them. It is not possible to review in detail even the many important studies that have shed new light on the essential problems since the publication of *The Origin of Australians* (Kirk & Thorne, 1976) and *The First Americans* (Laughlin & Harper, 1979).

In 1949 it was as important as it is today to consider Australians in a discussion of the American Indian (Birdsell, 1951). Typological analysis was then still in vogue and E. A. Hooton had encouraged the recognition of several diverse racial components in the analysis of American Indians (Howells, 1973*a*). Population genetics, particularly appropriate in blood group studies, rapidly displaced typological analyses (Boyd, 1939, 1950, 1951, 1963). The idea that Australoids or Amurians migrated as population entities or were carried in titrated solutions of genes into the New World had disappeared into the limbo of typology. Today we recognize that the Australians are scientifically important because they are a major population complex of considerable antiquity and were originally derived from the Asian mainland. As the Americans did later, they moved into a previously unoccupied continent and successfully dominated it. The basic similarities of these terminal populations in a larger continuum of populations are fully as important as the dissimilarities.

One of the first astute comparisons between these two areas is the remarkably prescient observation of J. B. S. Haldane presented in the Huxley Memorial lecture of 1956. Following a discussion of the adaptive value of skin pigmentation, he noted that we cannot argue that the very

14

dark pigmentation of Africans and Melanesians indicates a common ancestry. 'However, from the absence of a very dark indigenous race in tropical America we can suggest that the evolution of such a race requires over 10 000 years, and that the ancestors of Negroes and Melanesians have mainly lived in the tropics for more than that time' (Haldane, 1956, p. 4).

This comparison is a useful benchmark because of the large difference in degree of pigmentation and because of the accuracy of the time estimate. Ten thousand years agrees well with radiocarbon dates of valid context in South America (Turner & Bird, 1981) and also with the 12 000 year estimates for the continental United States where a large number of dates of determinable context have been analysed in rigorous detail (Haynes, 1969). Recent reanalysis using accelerator mass spectrometry to date those human skeletal materials previously dated between 15 000 and 70 000 BP have revealed major downward revisions so that no human remains in the New World are older than 11 000 ^{14}C years BP (Bada *et al.*, 1984; Taylor *et al.*, 1985). The 15 000 years we cite for the earliest Americans envisages 2000–3000 years of permanent occupation on the marine coast of the former Bering land bridge (Laughlin, 1967; Harper, 1980). Otherwise we should be obliged to retreat to 11 000–12 000 years based on single dates from archaeological sites in the Alaskan interior.

Our solution to the embarrassment of riches in the long record of scientific studies is to focus on three themes that singly and in combination bear upon (1) diversity, both within and between the areas under consideration, (2) time depth and nature of the entry route, particularly where this may have permitted or excluded multiple migrations, and (3) the evolutionary agenda, extending back to the middle Pleistocene of Asia, where continuities in several cranial and dental traits suggest that the traditional partition into fossil man and modern man rather than continuity is unnecessary and may have been seriously misleading.

Hunting – the unifying theme in human evolution

A part of the genetic repertoire of the populations under discussion, and the evolutionary agenda that led them to the places they now occupy, rests upon the early development of hunting as a way of life and its complex learning system (Spuhler, 1959; Laughlin, 1968). Hunting, in which we include scavenging, fishing, fowling and collecting, is the most ancient biobehavioural system known to humankind for it is the school of participant learning which shaped the human species. As a process it involves the systematic accumulation of vast amounts of knowledge on animal behaviour and its contexts, and also childhood instruction, inventiveness and problem solving.

The peopling of Australia and America demonstrates the migratory abilities of early hunters who successfully populated the last two major unoccupied areas in the world. Knowledge and practised skill are far more to the point in human migration than in any other single large animal migration, and these attributes were obviously possessed by the earliest migrants who crossed Wallace's Line and entered a radically different zoological world of monotremes and marsupials, and continuing common marine elements.

Following the Pleistocene achievement of the carnivorous–omnivorous diet made possible by hunting, including the manufacture of tools and the use of fire, the basis for food-sharing was extended into migratory abilities that uniquely conducted human populations to the far ends of the earth. Beyond this common human background there are important geographical differences in latitude and solar radiation, in periodic and restricted routes of entry, and in the annual and regional distribution of foods, all of which have significance for the migrants.

Migration into the Americas

Many prehistorians, influenced by Dawkin's (1866) claim that the Eskimo were the descendants of Upper Palaeolithic men of southern France and by Sollas (1924) who suggested a common Eskimo–Indian origin in the Upper Palaeolithic of Siberia, have expressed a belief that the earliest Americans followed reindeer/caribou, or mammoths, or some combination of megafauna. The term coined by de Laguna (1979, p. 17), 'following-the-reindeer theory', is an apt characterization of the tendency to attribute migration into the New World to the extreme attractiveness and numerical wealth of terrestrial animals, even during a glacial maximum when the most dependable resource base was undoubtedly the Bering Sea and its rivers, not the Arctic Circle. The argument for following the megafauna has usually involved an entry prior to the existence of the Bering land bridge (25 000–14 000) and sometimes the consumption of the mammoths and other megafauna as an explanation for their extinction. It is apparent that the basic evidentiary standards of scientific archaeology have often been suspended (Owen, 1984). In a well-searched examination of this topic, Bowdler (1977, pp. 231–2) has remarked: "It is ironic that man's role in extinction of Australian Pleistocene fauna is now being used to bolster the American argument'. Hunting people do not follow reindeer/caribou. They intercept them. Population fluctuation in caribou numbers and in the routes they follow vary from year to year. Consequently they are a welcome addition to an Eskimo economy but dependence upon them may result in dispersion or extinction (Burch, 1972).

The late Pleistocene migrants into the New World were limited to the southern coast of a land bridge which only appeared during a glacial period; the last such period, c 25 000 to 14 000 years ago, was of marked severity for land flora and fauna (Fig. 2.1). Further, the water temperatures limited human immersion to five or ten minutes, if followed by rapid rewarming. Fish, seals, walrus, beluga (white whales) and birds, also some caribou and an occasional musk-ox, were available in adequate numbers to sustain small isolates of well-clothed hunters. Recreating the early coastal environment depends not only on the zoological inventory of the Bering Sea, with special attention to those forms that prefer life

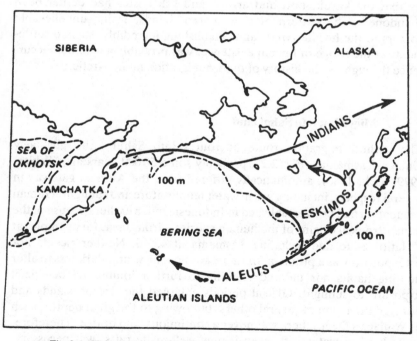

Fig. 2.1. The migration of the First Americans followed the southern coast of the now submerged Bering land bridge connecting Siberia and Alaska. This single small population bifurcated into two groups. The ancestral American Indians left the coast at the mouth of the Yukon River. The Bering Sea Mongoloid population remained on the coast and subsequently diverged into Aleuts and Eskimos. They still occupy the Arctic coasts of North America.

near the ice or breed only on land or ice fronts, but also upon latitudinally graded ethnographic analogy. Fortunately, the Eskimos inhabit a coastal area extending from 55° North to 80° North. Archaeological settlements of extinct Eskimos in Perry Land, the far north of Greenland, provide mute evidence of human ability to approach within a few hundred miles of the North Pole, though evidence as well that the cost of such hunting ability included extinction. These extinctions have been frequent in the Arctic.

The last Eskimo migration into Greenland of the 1860s, has been analysed by R. Pedersen (1962) and lists in tragic detail the challenges faced by a small band of some 40 migrants who had not only a severe environment to cope with but also their own disturbing antagonisms. The list of foods eaten included, on appropriate and desperate occasion, human corpses. The Polar Eskimo settlement of Thule was the intended goal of the migration, a settlement so isolated and dependent on winter-ice that the kayak, bow and arrow, and fish leister had earlier been abandoned. The pattern of tiny migrant bands, facing unbelievable horrors in the hostile Arctic and establishing incredibly isolated settlements at the fringe of human existence, was probably a common occurrence throughout the history of human migration in the Arctic.

Migration into Sahul land

The earliest migration route or routes into Sahul (Australia–New Guinea–Tasmania) (Fig. 2.2), probably some 50 000 years ago (White, 1979, Kirk, 1980), automatically differed from the Alaskan gateway in several respects, including warm water temperature and sunlight. Human occupation had been established in Indonesia much earlier because of the earlier tropical origins of the human species – from Java, for example, at 7° latitude, to and including Tasmania at 45° S. Neither rickets nor osteoporosis was a problem for there was no long winter darkness to alter hunting habits nor tailored clothing to further inhibit the beneficial exposure to sunlight. Glacial periods enlarged the size of islands and effected the union of several others, but owing to the great depths, such as Wallace's Deep, there was never a continuous land bridge. Therefore, some kind of watercraft, even if only well-made rafts, was necessary. Birdsell (1977) provides a detailed analysis of the possible migration routes and distances of water crossings from Sunda (now Indonesia) to Sahul.

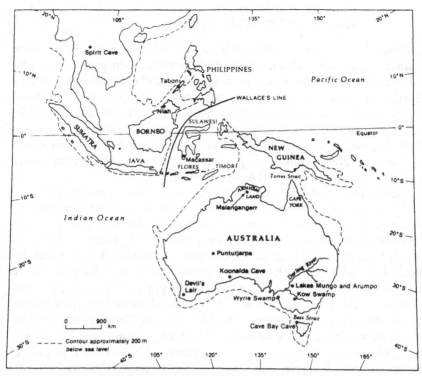

Fig. 2.2. The Sunda and Sahul shelf areas during the late Pleistocene show land areas formerly joined together. The former union of New Guinea, Australia and Tasmania (Sahul) remained separate from South-east Asia during the late Pleistocene. Therefore, some kind of watercraft was necessary for crossing the open water between Sunda and Sahul. (After White, 1979, p. 353.)

The relative sizes of the greater New Guinea–Australia–Tasmania territory and that of America are vastly different. Australia contains about five million square kilometres in contrast to the 25 million square kilometres of America. Assuming an additional one million square kilometres for Sahul at it glacial maximum, the ratio is somewhere around one to four. The north to south length from New Guinea to Tasmania is about 5000 kilometres, in contrast to some 9000 kilometres from northern North America to the South American Magellanic Islands. A significant amount of America is either high latitude or high altitude, and contains the smallest population densities in the high latitudes and much larger population densities in the high altitudes, where, unlike in the Arctic, food such as potatoes can be successfully grown.

Genetic differentiation in Sahul land

It appears that genetic and morphological diversity is greater in New Guinea, Australia and Tasmania than in all of America. There are considerable difficulties inherent in making sound comparisons between diverse areas and peoples, and in apportioning the variability in these diverse areas to time, size of founding populations, repeated or single migrations, host factors in disease resistance and other pertinent factors. Different measures must be employed for different bodies of data and consistency between different approaches must be carefully monitored, as Kirk & Thorne (1976, p. 6) have observed.

The most comprehensive and informative overview of genetic differentiation in Australia–New Guinea is provided by genetic distance analysis based on many marker genes and many populations. Fig. 2.3 is a dendogram for 19 New Guinea populations, one Island Melanesian population, and 12 Aboriginal populations in Australia (Keats, 1976). In discussing this study, Kirk (1979) notes that the most striking feature is the complete split between the Australian populations and the New Guinea–Melanesian populations. The genetic distance matrix on which this dendrogram is based also shows that the highland populations of Papua New Guinea and of Irian Jaya (western portion of the island) are most remote from five central desert populations in Australia, including the well-known Waljbiri and Aranda. Keats' analysis shows that the Melanesian-speaking groups, such as Motu, Fiji, Manus and Usiai, are more similar to northern Australians than to those in central Australia. This in turn suggests that the Melanesians were sea travellers, for which there is much other evidence, and that they left their genetic imprint on a broad front across northern Australia as well as on parts of coastal New Guinea (Keats, 1977). Unfortunately, there is no living Tasmanian population to complete the picture and here the osteological remains must be used for comparisons with Australia and New Guinea.

As a general consideration, there is a common assumption in many studies that the highland peoples of New Guinea represent earlier populations than those on the coast, unlike the situation in Australia where interior populations may have moved back to the coast at different times in their history.

A signal point to be underscored in using genetic distance estimates for the Pacific is that the number of allele frequencies employed often affects the outcome and creates strange bedfellows. A case in point is that of the Easter Islanders who, on the basis first of a few and then of as many as 13 alleles, showed some similarity to American Indians, and thus supported

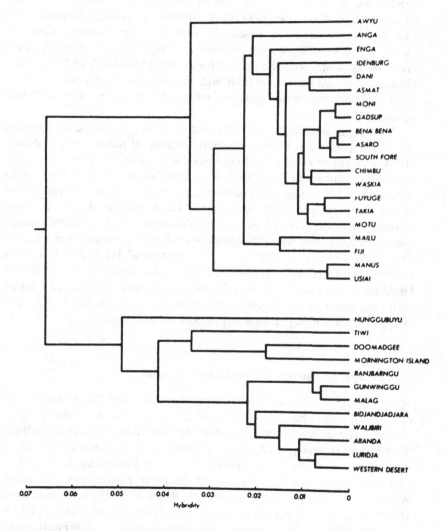

Fig. 2.3. Genetic separation of Australia, New Guinea and Island Melanesia is shown in the genetic distance dendrogram (Keats, 1976) employing 20 Melanesian and 12 Australian populations.

an imaginary migratory movement from the west coast of America into the Pacific. 'Results from the genetic distance analysis based on 28 blood genetic loci indicate, however, that the American Indians cluster more closely with the Japanese than they do with other Pacific populations, and that the Polynesians are genetically closest to the southern Chinese' (Kirk, 1979, p. 232). In this respect it is significant that the Central Australians cluster with the Fore of the Eastern Highlands of Papua New Guinea (also well known for their high frequency of Kuru), and together form a distinct population grouping when a fuller complement of 28 loci are employed.

The complicated but highly informative HLA gene complex provides important confirmation of the genetic history of Sahul. Sergeantson's (1984) analysis of HLA gene frequency distributions in 17 circum-Pacific populations finds that Australian Aborigines do not cluster closely with Melanesian, Micronesian or Polynesian populations, because Australians have a comparatively restricted range of HLA antigens. However, from a broad perspective, the HLA system reflects biological interrelationships which are congruent with other lines of evidence. A summary is given by Sergeantson (1984, p. 170): '. . . the Australoid HLA profile is very restricted and has been overlaid by Papuan elements in the New Guinea Highlands. Australia and New Guinea must have maintained contact with each other until at least 10 000 years ago since the HLA–A11.B40 linkage relationships persist today in both populations'.

Skeletal biology and prehistory of Sahul

Relationships between New Guinea, Australia and Tasmania can be considered with the use of skeletal evidence and its archaeological context. The skeletal picture provided in a preliminary study of multiple discriminatory analysis of linear dimensions of nine groups including Tasmania, Australia, New Guinea and New Britain by Giles (1976) identifies an interesting relationship between Tasmanians and north Australians of the Torres Straits Islands. While generally confirmatory of Howells' massive world-wide survey (1973a) and his distillation of 1976 (Howells, 1976), there are some interesting departures. After noting that the Philippine, Guam and Hawaiian series are all well away from Australians, Howells turns to eight small series of Pacific skulls, finding: 'This body of evidence . . . suggests that Tasmanians, Tolais (of the Gazelle Peninsula, New Britain) and Island Melanesians form a single constellation of populations, from which the south Australians, at least, are somewhat removed; at the same time, on a world-wide basis, the Australians still belong in the same major complex, and are in no reasonable way to be connected with Africans or archaic whites. Thus, . . . the Australians form one pattern, the Papuo-Melanesians another, but both fall in the same population complex' (Howells, 1976, p. 155). Thus, it is possible to use crania to support the genetic marker relationships and relate the now extinct Tasmanians to their larger population complex without resorting to exotic origins from either Africa or the Ainu of Hokkaido.

Though Australia and New Guinea shared a common population from some 50 000 years ago to the formation of Torres Strait that separated New Guinea from Australia 8000 to 10 000 years ago, there are now measurable differences in many traits. Many more data, especially skeletal remains, are needed to estimate how much and how many of the differences are consequent upon the separation of Australia and New Guinea and how many are due to more recent arrivals. The contextual data of the last few thousand years are remarkably clear. Torres Strait separates the agriculturalists of New Guinea, and the Pacific generally, from the hunters of Australia. The important taro plant appears in New Guinea about 7000 BC and the pig in New Guinea highland sites by 8000 BC. Both the plant and the pig owe their origins to South-east Asia and therefore indicate human migration in some degree, however limited, as well as trade relations. A probable population figure of the eighteenth century might be estimated at 250 000 for Australia in contrast to some 2 million for New Guinea (Howells, 1973b). The difference is clearly the result of agriculture and there is a close parallel in Central America and portions of South America where agriculture provided the base for very large population numbers. Indian corn, potatoes, squash and beans appear to have been independently developed from native plants in America, but the consequences are similar.

Two major archaeological findings provide an insight into the problem of variation within Australia well prior to the separation of Australia and New Guinea. These are the skeletal remains from Kow Swamp and Lake Mungo in south-eastern Australia. The signal fact is that they are different from each other, and to a lesser extent different from living Australians. The earlier remains from Lake Mungo, probably 30 000 years in age, are lightly built, gracile, with thin cranial vault bones and an overall delicacy unexpected in Australia (Howells, 1973a; Bowler & Thorne, 1976; Thorne, 1977; Thorne & Wolpoff, 1981). The more recent Kow Swamp series approaches the modern Australian Aborigines in being much more massive, with thick cranial vault bones, flat narrow forehead, large brow ridges and massive jaws without projecting chin.

These important finds pose many problems and admit several interpretations. They raise the spectre of multiple migrations from somewhere else on the one hand, and of greater variation in the original peopling of Australia and more regional evolution on the other hand. They also bring to mind the diversity present in the Upper Cave of Choukoutien of some 18 000 years ago. Until larger series of skeletons are found, and the full range of studies completed, including the discontinuous traits and the dental morphology, there can be satisfactory but not conclusive explanations of this diversity. The differences between the two provide encouragement to give more attention to rates of evolution within our single species in all its regional manifestations.

The thrust of Bowdler's (1977) analysis of the coastal colonization of Australia directs attention to the possible genetic consequences of a long period of coastal adaptation, including many movements around the coast, and then movement up the major river systems. As with the Bering Sea, the 18 000 B P glacial maximum was the time of lowest sea levels. The problem of locating archaeological sites is also identical: rising sea levels submerged all sites that were not protected by unique locations, on high promontories, on elevated strands or located up the rivers. As with the Greenland Eskimo model, it is most likely that the earliest people moved in both clockwise and counter-clockwise directions around Australia, but they were also able to migrate into the interior in the absence of a permanent ice-cap. Consequently, they could preserve and generate considerable variation.

Peopling the New World

It is far simpler to understand the peopling of the New World than that of Australia–New Guinea and the Pacific because of the stringent bottleneck entry during the severe glacial times of 15 000 years ago, the northern latitude, very cold water/ice transformation zone, the absence of early or fossil man, and very few population sources for a very small initial migration. Alaska is the gateway to the New World and, therefore, evolutionary events in Alaska configure the entire population history of the New World (Harper & Laughlin, 1982). To these factors must be added the overall homogeneity of the New World when compared with a similar area of Europe, Africa or Asia. This has been succinctly phrased by Turner who has systematically examined the morphology of more dentitions of people around the world than any other single investigator: 'There is in fact more dental variation between Aleuts and Eskimos, generally accepted genetic and linguistic relatives, than in the entire Indian population of North and South America' (Turner, 1983, p. 152).

A key question for the population prehistory of America is whether it consisted of one population making one migration and following one route, at one time, from Siberia to Alaska, or multiple migrations of the same or different populations over different routes, at different times (Williams, *et al.*, 1985). Phrased differently, did the diversity in America develop in Asia and transfer to America by migration, or did it develop in the 15 000 years since an initial migration? From the standpoint of population genetics, is the amount and patterning of diversity consonant with a single population origin, and if so, is there a positive indication of what in fact did happen? A part of our suggested scenario rests upon the quality and context of genetic and morphological data, a part rests upon the limited options stipulated in the geographic distribution of the relevant populations living in Alaska, a part rests upon the within-group and between-group analysis of sequenced populations, and a part rests upon multiple radiocarbon dates. One underlying assumption in the overall interpretation is simply that the greatest diversity is found in the oldest occupied area (Sapir, 1916), i.e. in Alaska rather than the subsequently occupied areas to the south. A second assumption is that population divergence precedes linguistic divergence, as for example with the Apache and Navaho of the American south-west, who retain their Athabascan language intact even though they have migrated several thousands of kilometres south of their northern homeland.

The nature of the land bridge

The first and most important geographic stipulation is the former Bering land bridge or platform which existed in some form between about 25 000 and 14 000 years ago when it was first breached by a marine canal (Hopkins, 1979). The inundation proceeded fairly rapidly so that the general coastline configuration of today was well indicated by 10 000 years ago, and contemporary sea levels were established some 6 000–5 000 years ago. The Bering platform reached its maximum southern extent at the time of maximum glaciation, 18 000 years ago. Bering Strait was not the focal point for the early prehistory of the Americans, even after it came into existence. The earliest coastal occupation is submerged, as it is in Sahul, but some descendants are still on the present coast.

The coastal configuration at its maximum extended from the Anadyr region of Siberia south-eastward toward Bristol Bay in south-western Alaska and westward again to its terminus at Umnak Island. Appropriately, lemmings inhabit Umnak and islands to the east. In spite of their reputed proclivity for periodically hurling themselves from cliffs into the sea, they are not good swimmers and confine their migrations to solid land. The surface topography of the platform was flat and nearly feature-less. Only eight small islands, such as St Lawrence, St Paul and St George, rose high enough to provide relief in an otherwise monotonous landscape.

The northern side of this Bering–Chukchi platform terminated in permanently ice-covered sea where only polar bears and seals could possibly have survived in small numbers. The vegetation of the platform was tundra, sparse and scattered, and in the northern region of present-day Bering Strait it was polar or arctic desert for the region was cold, dry and windy. The picture is well supported by pollen percentages, geomorphology and latitude (or sun angle). The reconstruction of the vegetation has been searchingly studied over a long period of time (Colinvaux, 1967; Ritchie & Cwynar, 1982; Colinvaux & West, 1984). In spite of the evidence for a frigid polar desert, a concept of a 'mammoth-steppe biome', containing megafauna articulated in a complex grazing ecosystem of horses, bison, musk-ox, caribou, antelope, moose and mammoths, became popular between 1965 and 1982. The actual finds of skeletons have been very modest; no example of coexistence of these herds has been brought forward and the botanical evidence indicates that they did not exist. This has been incisively phrased: 'As a corollary of these conclusions we suggest that the "arctic-steppe biome" never existed in the northern Yukon (nor in Alaska) during the most recent stadial–interstadial cycle (25 000 to 0 BP), but this position does not pose a serious paleoecological impasse since, as it happens, there is so far no evidence from that period of large numbers of either species or individuals of herbivorous vertebrates in the far northwest of the continent' (Ritchie & Cwynar, 1982, p. 126). There is some evidence that conditions were better in eastern Siberia 35 000 years ago, but then there was no Bering platform and, most probably, no people either.

Conditions on the Bering platform strongly suggest that there was no hunter's paradise in the interior. The only way across was to follow the habitable southern marine coast where the fish, seals and walrus, and other coastal resources such as birds and reindeer, were found in sufficient abundance to nurture a small population of some 300 persons, and rivers could be accessed as well for anadromous fish and seals that follow them. Therefore, we eliminate or place in suspense account the interior route that was earlier postulated along with the accepted coastal route (Laughlin, 1967). The official crossing of the platform occurred when the migrants stepped over a prehistoric international dateline, approximately in its present position, in their progress from the mouth of the Anadyr River to the mouth of the Yukon River, which earlier emptied into the Bering Sea south of St Lawrence Island and only a few hundred kilometres from the Anadyr River. It is likely that the people lived permanently on the coast for some period of time encompassing 1000 to possibly 3000 years and by moving their communities to adjust to the submerging coastline were delivered to the present coastline, still adhering to their coastal-marine economy. The exception were those few who ascended the Yukon River and slowly diverged to become Indians.

Genetic prehistory

A genetic time clock was prepared by using the gene frequencies of six loci of serological traits for all the Aleut, Eskimo and Athabascan groups of Alaska (Harper, 1980). A key point in the procedure followed here is the use of the natural hierarchy of ever more inclusive population units. North American natives comprise the most encompassing category for all 13 groups sampled are included, i.e. eight Athabascan tribes, one Aleut, three Yupik-speaking Eskimos (North; Siberian, Central and South; and Sugpiak) and one Inupiaq Eskimo (northern Eskimos of Alaska); those of Canada and Greenland are in the same recent linguistic division. The Aleuts and Eskimos are termed Bering Sea Mongoloids because they originated on the Bering Sea coast and they still live there, although they have also migrated into other coastal areas such as Kodiak Island, the Chukotka Peninsula of the USSR, Canada and Greenland. Their geographic positions appear in the linguistic map (Fig. 2.4) of Krauss (1975).

Using the gene frequency data provided by Scott (1979) for the 13 groups, the between-group divergence and the within-group divergence were calculated. Assuming that the evolutionary forces operating at each locus are relatively equal for each population over long periods of time, for as much as 19 000 years in this case, then the ratios of between-group distances are proportional to the interval of time since the bifurcation of each group, and the ratios of the within-group differences are proportional to the amount of time each population has had to accumulate genetic variation since its origin. Therefore, within-group diversity is equated with origin, and between-group variation is equated with bifurcation or divergence. The possibility that groups may begin formation (aggregate or originate) before they actually separate or migrate must be recognized. The separation time could be preceded by a period of aggregation or gestation.

Finally, the ratios were converted into temporal units by inserting an accurate starting date – a date for the initial divergence of Aleuts and Eskimos of 9000 years, based on radiocarbon dates from Umnak Island for Aleuts and from the Alaska Peninsula for Eskimos. This simply

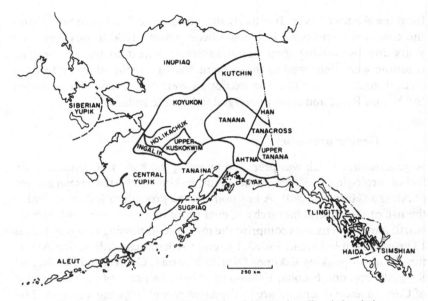

Fig. 2.4. Native peoples and languages of Alaska show linguistic relations
between the contiguous populations discussed in the text: Aleuts, Eskimos and
Athabascan Indians. The Eskimos are divided into two major divisions: the
Inupiaq of the north and the three Yupik (Siberian, Central and Pacific Coast
Sugpiak). The Athabascan dialect groups sampled are: Kutchin, Koyukon,
Tanana, Upper Tanana, Ahtna, Tanaina, Upper Kuskokwim and Ingalik. (After
Krauss, 1975.)

analysed procedure results in the genetic time clock seen in Fig. 2.5. Thus,
the genetic diversity within and between groups reflects the evolutionary
history of the populations within the comparison matrix.

We see that the between-group divergence of Athabascan and Bering
Sea Mongoloids is large and approximately equidistant for each group of
Bering Sea Mongoloids. The separation is therefore an old one and
occurred before Bering Sea Mongoloids diverged into Aleuts and
Eskimos. The Bering Sea Mongoloids in fact also display a gradient
sequence or cline from south to north. If the time of divergence between
Aleut and Eskimos is 9000 years, then the time of separation of Athabascan Indians and Bering Sea Mongoloids is secured by dividing 9000 years
by the ratio of Aleut/Eskimo divergence to Athabascan/Bering Sea
Mongoloid divergence, or $9000/(0.197/0.328) = 14\,985$ years. The
divergence between Yupik and Inupiaq is much smaller and therefore
must have occurred after the separation of Aleuts and Eskimos. The time
estimate is secured from $9000/(0.197/0.113) = 5162$ years before present.

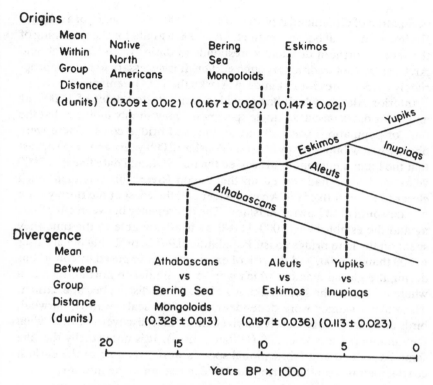

Fig. 2.5. The successive bifurcations of the First Americans established a genetic time clock. Bering Sea Mongoloids and Athabascan Indians separated 15 000 years ago, ancestral Aleuts and Eskimos separated 9000 years ago, and Inupiaq and Yupik Eskimos separated 5000 years ago. A comparison of the genetic time clock with glottochronological estimates of linguistic differentiation shows clearly that population divergence precedes linguistic divergence.

The within-group divergence values provide the basis for estimating time of origin. The values ascend with inclusiveness of the groupings and in themselves indicate the relative time of origin. This is converted into years for the origin of the population that gave rise to all native Americans (Aleuts, Eskimos and American Indians) by using the ratio of diversity within Eskimos to the diversity within all Alaskan populations.

Congruence between genetic prehistory and archaeology

A variety of gateways and checkpoints adds to the compatibility of the archaeological picture and the genetic picture. The earliest artefactual occupation of Greenland is placed at some 4000 BP, evidence of a migrant Eskimo group that became extinct. Our 5000 BP date for the beginning of the recent northern division corresponds to dating of the Denbigh and Arctic small tool tradition, though earlier. It is necessary to have appropriately earlier dates for Eskimos in Alaska than in Greenland.

Interior Alaska, which has provided well-attested dates of 11 000 BP and older dates based on single specimens, may simply indicate that the bulk of populations was still out on the land bridge coast. There were obviously few humáns anywhere in Alaska 12 000 years ago. We suggest that the human migrants had crossed the international date line by 15 000 years, and some had started up the Yukon River; this was then much closer to the mouth of the Anadyr River and the coast of the Bering shelf was far south of St Lawrence Island. The discrepancy between our 15 000 BP and the established 12 000–11 000 BP is attributable to the time they spent on the land bridge coast. Population size was probably much closer to 300 than to 3000, and the risk of extinction was imposed by conditions during the glacial maxima which were most similar to north Greenland where all except the Polar Eskimos of the Thule district became extinct. The Polar Eskimos were dependent upon fish, seals, walrus, narwhal, birds and caribou and barely survived until their discovery in 1818 when they numbered less than 200 (Gilberg, 1976). It is noteworthy that the same basic food and fabricational species were available to the earliest coastal migrants on the Bering land bridge but not in the interior.

Genetic and morphological differentiation in the New World

Using gene frequencies of red cell antigens in measuring the genetic affinities of 53 North American Indian tribes, Spuhler found a significant but not high correlation among biology, language and culture. In general, gene frequency affinities were a little stronger by language family than by culture area: the genetic distances classified languages into the families in 64.7% and tribes into their culture areas in only 58.5% (Spuhler, 1979). This meaningful result is consonant with various studies where the focus is on the Aleut–Eskimo stock and the contiguous Athabascans (Harper, 1980), and where the focus is primarily on the Aleut–Eskimo stock (Ferrell *et al.*, 1981).

Thus, the Aleut–Eskimo linguistic unity and distinction from all others has been apparent from many useful studies from 1916 to the present. No related languages have been demonstrated in Siberia, and there seems to be a remarkable paucity or absence of loan words from Athabascan into Eskimo (Bergsland, 1979; Krauss, 1983). The latter indicates the isolation of the Aleut–Eskimo, and in fact raises the question of where the Athabascans lived in the last 10 000 years.

Within the Aleut–Eskimo population system there is clear evidence that the Inupiaq language developed from the central or northern Yupik language, and that the separation between south, central and northern Yupik languages (or major dialect groups) is much older than the separation between Yupik and Inupiaq. Inside the Aleut–Eskimo system the congruence between metrical and non-metrical cranial variation and linguistic categorization has been handsomely demonstrated (Zegura, 1975, 1978). The presence of cranial clines, especially obvious in the cranial index, has been observed by Hrdlička (1930) and by Laughlin, Jørgensen & Frøhlich (1979). The distinctiveness of the mandible and its similarity between that of the Aleutians and Greenland has been massively demonstrated by Frøhlich (1979), and the dental morphology has been presented in detail by Moorrees (1957), Pedersen (1949) and Turner (1967, 1983).

All the facts support the interpretation of Spuhler: 'On a world scale, the North American Indians, Eskimos and Aleuts are a genetically distinct race or breeding population, related most closely to the Mongoloid peoples of eastern Asia' (Spuhler, 1979, p. 177). The same conclusion has been reached by others: 'The Eskimo/Amerindian cluster is clearly separate from the aboriginal northern Siberian populations from the Taimir peninsula' (Ferrell *et al.*, 1981, p. 357).

The best explanation is that of the development of all the American groups from a single small migration restricted to the southern habitable coast of the Bering land bridge under conditions that severely limited population survival, let alone expansion. The Aleut–Eskimo have often been accorded a separate and recent arrival by ignoring the genetic differentiation within the sequenced divisions and by assuming a homogeneity which does not now and never did exist. A case in point is blood type B. It has been attributed to recent introduction into the Aleut–Eskimo and to a postulated recent arrival of the Aleut–Eskimo. However, its distribution demonstrates that it appeared at least 10 000 years ago before the bifurcation between the Aleuts and the Eskimos because it is found all the way from the Aleutians to Greenland. It achieves its highest value in the small Anaktuvuk Eskimo isolate as a result of the contributions of one man who endowed his children and grandchildren with the gene (Laughlin, 1957). It drops to near zero in some of the small and isolated groups of north-eastern Canada and Greenland (Heinbecker & Pauli, 1927). A related sampling effect is seen in a small group of Aleuts and in a small group of Eskimos in which all of the genes for N are carried in heterozygous MN. Thus, the presence of B among the Aleut–Eskimo and its absence among the Indians is one useful measure of group integrity. There has been little gene exchange between Eskimos and Indians since diverging.

It is most likely that the original migrant population of 15 000 years ago carried the B gene and that by sampling accident the small group who ascended the Yukon River and subsequently isolated themselves geographically from the coastal population did not carry this low frequency gene. Neither the Aleut–Eskimo stock nor the gene for B is recent. The northern Inupiaq division of Eskimos is recent, i.e. 5000 years in its formation and separation from the older Bering Sea coast Yupik Eskimos.

Thus, the interpretation of a single gene can be edifying or misleading, depending on our knowledge of the population structure and its time depth, which differs in different parts of its present domain. Equally to the point is the advantage of using several genes; the single genes become more informative when employed in a larger multivariate analysis. Single gene analysis would commit the New World analysis to a layer-cake model, having three layers consisting of one population with B, A and O, one with A and O, and the third with only O. Counterpart populations in eastern Asia and in Siberia cannot be found, and if they did exist it would be evidence of the absence of evolution. It appears that the diversity in Alaska has developed because of the first arrival of the migrant population there, which was because it was and has been the only gateway into the New World. As the encroachment of the Bering Sea took effect, part of the coastal population continued their marine economic adaptation and extended their occupation of this linear domain.

The single species perspective

The facts, as nearly as they can be represented by the combination of fossil human remains and living populations, suggest that there has been but one human species from the Pleistocene to the present. Early differentiation was as natural in the middle Pleistocene as it has been in the last 15 000 years. Consequently, it is possible to follow continuous lines of human evolution and differentiation from southern Asia into Australia and from northern Asia into the Americas. This view eliminates any particular Rubicon that was crossed by modern humans when they left their 'primitive' morphologies on the near bank and crossed over to the modern bank. All living populations are products of earlier evolution and all retain various features, of which many are useful for tracing their evolutionary lineages to the regions in which some characterization of contributing populations has been made possible by the discovery and analysis of earlier parental populations. The dozen or more characters used to trace the American Indians back to north China and, equally, the various characters used to trace Australian Aborigines back to South-east Asia does not imply in any way that the original occupants of Australia or of the New World are 'primitive'. Rather, they indicate that human origins can be assessed in a scientific fashion and provide the hope that an increasingly improved knowledge of the rates of human evolution, and the contextual factors that must be better known, will enhance the human condition now and more so in the future.

The best exposition of the single species thesis is contained in the study of 'The skull of *Sinanthropus pekinensis*' by Weidenreich (1943). Its relevance is all the more impressive because of his discussions of the Australians and also the Aleuts, Eskimos and Indians. Weidenreich makes abundantly clear the single species nature of all hominids by varied reiteration. 'But concede this also, it would not be correct to call our fossil *Homo pekinensis* or *Homo erectus pekinensis*: it would be best to call it *Homo sapiens erectus pekinensis*. Otherwise it would appear as a proper "species", different from *Homo sapiens* which remains doubtful to say the least' (Weidenreich, 1943, p. 246). He continues (p. 256): 'The hominids have formed – and still form – one family, in a strictly taxonomic sense, our species are all more or less related to each other in spite of manifold regional variations'.

According to Weidenreich's terminology, *Pithecanthropus* and *Sinanthropus* are classed as two racial, not specific, variants of early man. 'There is then an almost continuous line leading from *Pithecanthropus* through *Homo soloensis* and fossil Australian forms to certain modern primitive Australian types' (Weidenreich, 1943, p. 276).

Even if the 12 traits nominated by Weidenreich for support of the concept of a line leading from *Sinanthropus* to modern Mongoloids is reduced to the four most objectively defined traits from different locations on the skull, we have tangible evidence to work with. It is informative to see the frequency and distribution of these traits in the New World, and to a much lesser extent in Polynesians. These are the Inca bone of the posterior cranial vault, exostoses of the auditory meatus at the base of the cranium, the mandibular torus and shovel-shaped incisors, all developmentally independent of each other. The Inca bone is obviously of notable frequency in Peruvian crania, but it also has a generally high frequency in most Mongoloid-derived groups, including Aleuts and

Eskimos. Exostoses of the auditory meatus achieve high frequencies in Indians but are absent in Aleuts and Eskimos (Stewart, 1979). The mandibular torus is of very high frequency in Aleuts and Eskimos, but notably lower in Indians, with some evidence of a latitudinally influenced frequency distribution. Shovel-shaped maxillary incisors are of high frequency in all New World inhabitants and in Asiatic Mongoloids as well. The distribution of this trait is one of the most informative single trait distributions in human variation (Fig. 2.6; Turner, 1983). Though not yet proved at a high level of testing, the case for a line from *Sinanthropus* to modern Mongoloids looks very promising.

Migration north of Peking has been much more recent than migration south, and the recentness of the northern Inupiaq division of the Eskimos into the Arctic is an indication of the biological rather than cultural inhibitions on migration into the real Arctic. Following the Lower Cave finds of Choukoutien, a small series of skeletons from the Upper Cave was found, now dated at about 18 000 years ago. The evaluation of these

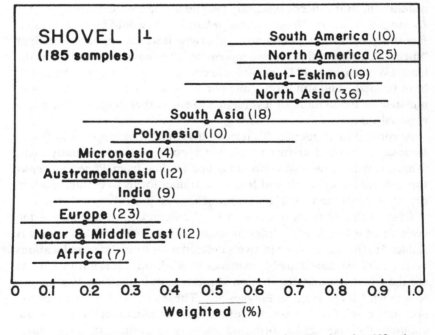

Fig. 2.6. The frequency of upper central incisor shoveling, based on 185 series, shows the close relationship of North Asia and America, and the lower frequencies among the peoples of Sunda, Sahul and the Pacific. (After Turner, 1983, p. 152.)

skulls contributes to an understanding of the evolution of Mongoloids and of their migration into the New World.

The three skulls became famous for the fact that they did not closely resemble each other and could be typologically compared with Melanesians, Eskimos and Europeans (Weidenreich, 1939). Discussing them in 1943 (p. 251), Weidenreich nominates the 'Old Man' of Choukoutien as an early East Asiatic type and, because some prehistoric Mongols reveal similar characters, as a proto-Mongoloid. Reviewing the two female skulls, one that looks like a Melanesian of today and the other like an Eskimo, he pronounces the essential evaluation: 'But to identify them with those races would not be justified. Their designation as "Melanoid" and "Eskimoid" means merely that those features we consider as typical of those races were already developed in the Far Eastern Palaeolithic man of late Magdalenian' (Weidenreich, p. 255). Thus, the designation of 'unmigrated American Indians' bestowed by Howells (1959, p. 300) is apt and possibly prophetic.

Conclusions

Australia–New Guinea (Sahul) was early occupied by a highly skilled group of hunters some 50 000 years ago, who travelled by water from Indonesia. Their first adaptation was primarily coastal, though much of the early evidence has been submerged by rising sea level following the glacial maximum of 18 000 years ago. The next major event which affected the population was the separation of New Guinea and Australia (and also Tasmania) some 10 000 years ago. Some of the diversity between Australia and New Guinea has developed only in the last 10 000 years and has included the addition of traits from Indonesia which have been increasingly Mongoloid.

The diversity within Australia is all the more interesting because of the several different systems involved (hair colour, form and distribution, pigmentation, as well as in blood groups, serum proteins and red cell enzyme systems), and because of the smaller size of Australia compared with the New World. Within Australia there are tantalizing hints that there is greater differentiation in the north than to the south (excepting Tasmania) which parallels the situation in the New World where the greatest diversification is found in Alaska rather than farther south.

The American Indians, including the Aleut–Eskimo stock, form a geographic race or population complex. Although separated from eastern Asia for some 15 000 years, its affinities with the Mongoloid populations of eastern Asia are well demonstrated in blood genetic markers, dental

36 *W. S. Laughlin & A. B. Harper*

morphology, metrical and non-metrical traits of the skeleton, as well as in the straight black hair and moderate pigmentation. The relative homogeneity of the people of this continent is adequately explained by a single migration along the habitable southern marine coast of the former Bering land bridge prior to the formation of Bering Strait. Geographic isolation, and small population sizes, have been the chief factors in distributing diversity in the New World. The gene for blood type B is not a recent gene in the New World and the Aleut–Eskimo complex is not a recent population but rather is one of several that has developed from the original migrant population of 15 000 years ago.

Australia and the New World are related through the regional racial, not specific (subspecific), evolution that is presumably traceable to *Homo erectus pithecanthropus* and *Homo erectus sinanthropus* of Java and North China respectively. There is presumptive evidence for continuity from 'Peking Man' to contemporary Mongoloids, and for continuity from 'Java Man' to Australian–New Guinea Aborigines. The greater variability that exists in Australia–New Guinea than in the New World reflects the much greater time depth and, to a limited extent, subsequent migrations or gene flow, primarily within the last 10 000 years since the separation of Sahul into Australia and New Guinea. The presence of the gene for blood group B in northern Australia may well have been established there by the first immigrants. This possibility has been earlier recognized: 'But the presence of these New Guinea markers in Arnhem land may merely reflect a common substratum of population dating to before the physical separation of New Guinea and Australia, some 8000 or more years ago' (Balakrishnan, Sanghvi & Kirk, 1975, p. 102). Multivariate studies of Australia–New Guinea, and those of North America and South America, provide a basis for tracing clines in a variety of traits, such as the very high N frequency in Australia and the very low N frequency in the entire New World. They interestingly suggest that the gene for B is a very old gene in eastern Asia as well as in India and Africa and that it may have reached Australia earlier than 30 000 years ago when the first migration occurred there, as well as in America. There are no profound differences within or between Australia–New Guinea and the New World.

References

Bada, J. L., Gillespie, R., Gowlett, J. A. J. & Hedges, R. E. M. (1984). Accelerator mass spectrometry radiocarbon ages of amino acid extracts from California Palaeoindian skeletons. *Nature*, **312**, 442–4.

Balakrishnan, V., Sanghvi L. D. & Kirk, R. L. (1975). *Genetic Diversity among Australian Aborigines*. Canberra: Australian Institute of Aboriginal Studies.
Bergsland, K. (1979). The comparison of Eskimo–Aleut and Uralic. *Fenno-Ugrica Suecana*, 2, 7–18.
Birdsell, J. B. (1951). The problem of the early peopling of the Americas as viewed from Asia. In *Papers on the Physical Anthropology of the American Indian*, editor. W. S. Laughlin, pp. 1–68. New York: Viking Fund.
Birdsell, J. B. (1977). The recalibration of a paradigm for the first peopling of Greater Australia. In *Sunda and Sahul: Prehistoric Studies in Southeast Asia, Melanesia and Australia*, ed. J. Allen, J. Golson & R. Jones, pp. 113–67. London: Academic Press.
Boyd, W. C. (1939). Blood groups. *Tabulae Biologicae*, 17, 113–240.
Boyd, W. C. (1950). *Genetics and the Races of Man*. Boston: Little, Brown.
Boyd, W. C. (1951). The blood groups and types. In *The Physical Anthropology of the American Indian*, ed. W. S. Laughlin, pp. 127–37. New York: Viking Fund.
Boyd, W. C. (1963). Four achievements of the genetical method in physical anthropology. *American Anthropologist*, 65, 243–52.
Bowdler, S. (1977). The coastal colonization of Australia, In *Sunda and Sahul: Prehistoric Studies in Southeast Asia, Melanesia and Australia*. ed. J. Allen, J. Golson & R. Jones, pp. 205–46. London: Academic Press.
Bowler, J. M. & Thorne, A. G. (1976). Human remains from Lake Mungo: discovery and excavation of Lake Mungo III. In *The Origin of Australians*, ed. R. L. Kirk & A. G. Thorne, pp. 127–38. Canberra: Australian Institute of Aboriginal Studies.
Burch, E. A. (1972). The caribou/wild reindeer as a human resource. *American Antiquity*, 37, 339–68.
Colinvaux, P. A. (1967). Bering land bridge: Pollen evidence for spruce in late land bridge times. *Science*, 156, 380–3.
Colinvaux, P. A. & West, F. H. (1984). The Beringian ecosystem. *The Quarterly Review of Archaeology*, 5(3), 10–16.
de Laguna, F. (1979). Therkel Mathiassen and the beginnings of Eskimo Archaeology. In *Thule Eskimo Culture: An Anthropological Retrospective*, ed. A. P. McCartney, pp. 10–53. Archaeological Survey of Canada, Paper No. 88, Ottawa: Mercury series. National Museum of Man.
Dawkins, W. B. (1866). Eskimo in the south of Gaul. *Saturday Review*, London, December 8, 1866.
Ferrell, R. D., Chakraborty, R., Gershowitz, H., Laughlin, W. S. & Schull, W. J. (1981). The St Lawrence Island Eskimos: genetic variation and genetic distance. *American Journal of Physical Anthropology*, 55, 351–8.
Frøhlich, B. (1979). *The Aleut–Eskimo Mandible*. Ph.D. Thesis. Storrs: University of Connecticut.
Gilberg, R. (1976). The Polar Eskimo population, Thule District, North Greenland. *Meddelelser Om Gronland*, 203(3), 1–87.
Giles, E. (1976). Cranial variation in Australia and neighboring areas. In *The Origin of Australians*, ed. R. L. Kirk & A. G. Thorne, pp. 161–72. Canberra: Australian Institute of Aboriginal Studies.
Haldane, J. B. S. (1956). The argument from animals to men: An examination of

38   W. S. Laughlin & A. B. Harper

its validity for anthropology. *Journal of the Royal Anthropological Institute*, **86(2)**, 1–14.
Harper, A. B. (1980). Origins and divergence of Aleuts, Eskimos and Indians. *Annals of Human Biology*, 7, 547–54.
Harper, A. B. & Laughlin, W. S. (1982). Inquiries into the peopling of the New World: Development of ideas and recent advances. In *The History of Physical Anthropology, 1930–1980*, ed. F. Spencer, pp. 281–384. New York: Academic Press.
Haynes C. V. (1969). The earliest Americans. *Science*, **166**, 709–15.
Heinbecker, P. & Pauli, R. H. (1927). Blood grouping of the Polar Eskimo. *Journal of Immunology*, 13, 279–83.
Hopkins, D. M. (1979). Landscape and climate of Beringia during late Pleistocene and Holocene time. In *The First Americans: Origins, Affinities and Adaptations*, ed. W. S. Laughlin & A. B. Harper, pp. 15–41. New York: Gustav Fischer.
Howells, W. W. (1959). *Mankind in the Making*. New York: Doubleday.
Howells, W. W. (1973a). *Cranial Variation in Man. A Study in Multivariate Analysis*. Peabody Museum Papers **67**, 259 pp. Cambridge, Mass: Harvard University.
Howells, W. W. (1973b) *The Pacific Islanders*. London: Weidenfeld and Nichols.
Howells, W. W. (1976). Metrical analysis in the problem of Australian origins. In *The Origin of Australians*, ed. R. L. Kirk & A. G. Thorne, pp. 141–60. Canberra: Australian Institute of Aboriginal Studies.
Hrdlička, A. (1930). *Anthropological Survey in Alaska*. Washington, DC: US Government Printing Office.
Keats, B. (1976). *Genetic Aspects of Growth and of Population Structure in Indigenous Peoples of Australia and of New Guinea*. Ph.D. Thesis. Canberra: Australian National University.
Keats, B. (1977). Genetic structure of the indigenous populations in Australia and New Guinea. *Journal of Human Evolution*, 6, 319–39.
Kirk, R. L. (1979). Genetic differentiation in Australia and the Western Pacific and its bearing on the origin of the First Americans. In *The First Americans: Origins, Affinities and Adaptations*, ed. W. S. Laughlin & A. B. Harper. pp. 211–37. New York: Gustav Fischer.
Kirk, R. L. (1980). Population movements in the Southwest Pacific: The genetic evidence. In *Indonesia: Australian Perspectives*, ed. J. J. Fox *et al.* pp. 45–56. Canberra: Research School of Pacific Studies, Australian National University.
Kirk, R. L. & Thorne, A. G. (eds) (1976). *The Origin of Australians*. Canberra: Australian Institute of Aboriginal Studies.
Krauss, M. E. (1975). *Native Peoples and Languages of Alaska*. Fairbanks: Alaska Native Language Center, University of Alaska.
Krauss, M. E. (1983). Alaska native languages: Past, present, and future. In *Cultures of the Bering Sea Region: Papers from an International Symposium*, ed. H. N. Michael & J. W. VanStone, pp. 169–208. New York: International Research and Exchange Board.
Laughlin, W. S. (1957). Blood groups of the Anaktuvuk Eskimos. *Anthropological Papers of the University of Alaska*, 6, 5–15.

Laughlin, W. S. (1967). 'Human migration and permanent occupation in the Bering Sea area. In *The Bering Land Bridge*, ed. D. M. Hopkins, pp. 409–50. Stanford: Stanford University Press.

Laughlin, W. S. (1968). Hunting: An integrating biobehavior system and its evolutionary importance. In *Man the Hunter*, ed. R. B. Lee & I. Devore, pp. 304–20. Chicago: Aldine Publishing.

Laughlin, W. S. & Harper, A. B. (eds.) (1979). *The First Americans: Origins, Affinities and Adaptations*. New York: Gustav Fischer.

Laughlin, W. S., Jørgensen, J. B. & Frøhlich B. (1979). Aleuts and Eskimos: Survivors of the Bering land bridge. In *The First Americans: Origins, Affinities and Adaptations*, ed. W. S. Laughlin & A. B. Harper, pp. 91–103. New York: Gustav Fischer.

Moorrees, C. F. A. (1957). *The Aleut Dentition*. Cambridge, Mass. Harvard University Press.

Owen, R. C. (1984). The Americas: The case against an ice-age human population. In *The Origins of Modern Humans: A World Survey of the Fossil Evidence*, ed. F. H. Smith & F. Spencer, pp. 517–63. New York: Alan R. Liss.

Pedersen, R. (1962). The last Eskimo migration into Greenland. *Folk*, 4, 95–110.

Pedersen, P. O. (1949). The East Greenland Eskimo dentition: Numerical variation and anatomy. *Meddelelser om Gronland*, 142, 1–256.

Ritchie, J. C. & Cwynar, L. C. (1982). The late quaternary vegetation of the North Yukon. In *Paleoecology of Beringia*, ed. D. M. Hopkins, J. V. Matthews, Jr, C. E. Schweger & S. B. Young, pp. 113–26. New York: Academic Press.

Sapir, E. (1916). Time perspective in aboriginal American culture, a study in method. *Canada Department of Mines, Geological Survey, Memoir*, 90, 1–87.

Scott, E. M. (1979). Genetic diversity of Athabascan Indians. *Annals of Human Biology*, 6, 241–7.

Sergeantson, S. W. (1984). Migration and admixture in the Pacific. *Journal of Pacific History*, 19, 160–71.

Sollas, W. J. (1924). *Ancient Hunters and their Modern Representatives*, 3rd edn. London: Macmillan.

Spuhler, J. N. (1959). Somatic paths to culture. In *The Evolution of Man's Capacity for Culture*, ed. J. N. Spuhler, Detroit: Wayne State University Press.

Spuhler, J. N. (1979). Genetic distances, trees and maps of North American Indians. In *The First Americans: Origins, Affinities and Adaptations*, ed. W. S. Laughlin & A. B. Harper, pp. 135–83. New York: Gustav Fischer.

Stewart, T. D. (1979). Patterning of skeletal pathologies and epidemiology, In *The First Americans: Origins, Affinities and Adaptations*, ed. W. S. Laughlin & A. B. Harper, pp.257–74. New York: Gustav Fischer.

Taylor, R. E., Payen, L. A., Prior, C. A., Slota, P. J. Jr, Gillespie, R., Gowlett, J. A. J., Hedges, R. E. M., Jull, A. J. T., Zobel, T. H., Donahue, D. J. & Berger, R. (1985). Major revisions in the Pleistocene Age assignments for North American human skeletons by C-14 accelerator mass spectrometry: None older than 11 000 C-14 years B P. *American Antiquity*, 50(1) 136–40.

Thorne, A. G. (1977). Separation or reconciliation? Biological clues to the development of Australian society. In *Sunda and Sahul: Prehistoric Studies in Southeast Asia, Melanesia and Australia*, ed. J. Allen, J. Golson & R. Jones, pp. 187–204. London: Academic Press.

Thorne, A. G. & Wolpoff, M. H. (1981). Regional continuity in Australian Pleistocene hominid evolution. *American Journal of Physical Anthropology*, **55**, 337–49.

Turner, C. G. II (1967). *The Dentition of Arctic Peoples*. Ph.D. Thesis, University of Wisconsin.

Turner, C. G. II (1983). Dental evidence for the peopling of the Americas. In *Early Man in the New World*, ed. R. Shutler, pp. 147–57. Beverley Hills: Sage Publications.

Turner, C. G. II & Bird, J. (1981). Dentition of Chilean Paleo-Indians and peopling of the Americas. *Science*, **212**, 1053–5.

Weidenreich, F. (1939). On the earliest representatives of modern mankind recovered on the soil of East Asia. *Peking Natural History Bulletin*, **13(3)** 34–44.

Weidenreich, F. (1943). The skull of *Sinanthropus pekinensis*, *Palaeontologia Sinica*, New Series D, **10** 1–298.

White J. P. (1979) Melanesia, In *The Prehistory of Polynesia*, ed. J. D. Jennings, pp. 352–77. Cambridge, Mass.: Harvard University Press.

Williams, R. C., Steinberg, A. B., Gershowitz, H., Bennett, P. H., Knowler, W. C., Pettitt, D. J., Butler, W., Baird, R., Dowda-Rea, L., Burch, T. A., Morse, H. G., & Smith, C. G. (1985). Gm allotypes in Native Americans: Evidence for three distinct migrations across the Bering land bridge. *American Journal of Physical Anthropology*, **66**, 1–19.

Zegura, S. L. (1975). Taxonomic congruence in Eskimoid populations. *American Journal of Physical Anthropology*, **43**, 271–84.

Zegura, S. L. (1978). The Eskimo population system: Linguistic framework and skeletal remains, In *Eskimos of Northwestern Alaska*. ed. P. L. Jamison, Zegura, S. L., & Milan F. A. pp. 8–30. Stroudsburg, Penn.: Dowden, Hutchinson and Ross.

3 Migration in the recent past: societies with records

D. F. ROBERTS

The fact that a society maintains vital records for demographic purposes does not necessarily mean that there is a ready-to-hand source of data for the study of migration. Such records are usually collected for some other purpose, usually administrative. In a census, while the record sheets tell the location of a given individual on a particular date, these are rarely available to investigators (in the United Kingdom not for 100 years) and the data are often engulfed in the categorization required for demographic analysis. Direct questions on migration are considered of low priority, time-consuming to administer, and are often discarded when the number of questions to be asked is limited by cost. The location may be specified in some unit of space of little interest to the enquirer and, if large, migration within it will not be indicated. The boundaries of the space units may have changed between censuses, so that the same address appears in what seem to be two different localities in consecutive censuses. Use of fixed periods of time gives no indication on multiple moves and return moves during the interval; if a question is asked on absence from the census address for, say, a year, this will not identify the seasonal migrants. Census data can reveal little on the motivations for migration or the characteristics of those who migrate. Something may however be inferred from them. For example, if the age–sex pyramid in a population shows a pronounced deficit of young male adults, it may suggest absentee labour and therefore that migration, permanent or temporary, has occurred; or it may reflect heavy mortality in a recent war. In other words, utilization of data from a census for analysis of migration usually requires other relevant information as well.

Vital registrations of births, deaths and marriages, and their predecessors, the parish records of baptisms, marriages and funerals, are more useful. Such registers allow the identification of individuals, and by means of modern computer linking it is possible to compare the location of an individual at birth, marriage and death. However, the same location of an individual on all three occasions may or not represent lack of migration, for he may have returned to his birth area to marry or to die after wandering half around the world. Moreover, it frequently happens that

41

an individual appears in only one set of records, and cannot be traced before or afterwards; so for migrants the information available is often restricted. There are in some countries other types of register, for example in England and Wales the National Health Service Central Register, intended to monitor *per capita* payment of local doctors, and to which every transfer of a patient from one practice to another is notified. As a continuing source, it provides valuable information for studies of internal migration (e.g. Devis & Southworth, 1984, on the demographic features of migrants; Grundy, 1986, on the relationship of fertility to family movements). Occasionally small communities keep genealogical records, particularly new colonizing settlements or religious entities, which often give historical perspective to the migration inward or outward that has occurred.

This is not to say that useful material cannot be extracted from census, vital, and register records, but in asking questions of them the fact that they were collected for some purpose other than the assessment of migration must be borne constantly in mind. Most useful are sample surveys, which are on a smaller scale, for example in the United Kingdom the General Household Survey and the Labour Force Survey. These are especially useful if data in them can be linked to other types of record. The General Household Survey has covered annually about 15 000 private households in Great Britain since 1971. The Labour Force Survey provides statistics in each of the countries in the European community biennially. Similar surveys are available in many other countries which keep vital records.

Though everybody knows what is meant by migration – change in location of residence – in reality migration is quantitatively and qualitatively a highly heterogeneous process. It may concern a single individual who moves to seek employment or to marry; a group of individuals who are deported (as in the settlement of Australia) or recruited for labour (such as maintain the economy in South Africa) or for military service. It may concern a family or group of families who seek a better life elsewhere, or join their members who moved earlier as individuals (as in so much of the peopling of the New World). It may concern a subpopulation, particularly religious minority groups, such as the Huguenots or Jews, seeking religious freedom. Or indeed it may be an intrinsic feature of the way of life of a total population, as with the gypsies and the nomadic and hunter–gatherer peoples. For the individuals concerned it may be temporary or permanent, recurrent, seasonal, or occur once in a lifetime. For the populations concerned it may occur in a single generation or it may represent a continuing stream over several generations. It may be initiated by economic, social, political, psychological or religious pressures. it may be local (usually individual or family) or it may be distant. Its biological effects and implications for the population of origin, for the recipients, and for the migrants themselves, depend on its scale and its nature. Once arrived in a new locality, the migrants may retain their attitudes, religion, language, culture and genetic identity for a long period as with the Amish in the United States (McKusick, 1978), or there may be quite rapid breakdown and assimilation into the host population as with the Hungarians in Louisiana (Koertvelyessy, 1983).

This essay illustrates various ways in which data available in societies with records have been applied to illuminate the migration process and some of its biological effects.

Migration and demography

The amount of migration

From census and registration material can be derived useful estimates of various migration features. Sometimes the records allow direct estimates of migrant numbers, as in those who arrived in the New World after transatlantic migration; between 1815 and 1915 30 million people arrived in the United States, nearly two and a half million arriving in the 1850s alone, and the number who arrived between 1815 and 1860 was greater than the population of the whole country in 1790.

But the amount of migration can also be estimated less directly. Since population size only changes as a result of births, deaths and migration, if a total population is known at two dates, the number of births in the interval can be added to the first total, the deaths subtracted, and the difference of the sum from the second total indicates the amount of migration in or out. Baines (1972) applied this method to census data on county of birth and current residence to examine migration in Cornwall in 1861–1870. He estimated the average size of the male immigrant group in this period to be 15 364. According to the differential death rate accepted for migrants, the estimates of emigrants from the county ranged from 32 786 to 35 092. Of these, 9053–13 073 moved to other counties in England and Wales, and 22 019–23 722 moved abroad or to Ireland or Scotland.

Comparison of records of named individuals at two dates provides a second indirect estimate of the amount and pattern of migration. Villaflor & Sokoloff (1982) examined the colonial militia muster rolls, which

44   *D. F. Roberts*

contained both the place of birth and of residence of recruits at enlistment in the United States in the eighteenth century. At the time of the French and Indian War, in 1756–1763, recruits residing in New York but born overseas had come preponderantly from Ireland (40%), Germany (24%) and England and Wales (21%). Those enlisting in Pennsylvania showed a similar pattern, 63% from Ireland, 21% from Germany, and 10% from England and Wales. By contrast, Maryland had 36% from England and Wales, 25% from Ireland, and 11% from Germany, and Virginia showed a similar distribution having 45% from England and Wales, 43% from Ireland, but 17% from Scotland and only 2% from Germany. A generation later, at the time of the American Revolution (1775–1783), the pattern was generally the same, but there were fewer from Germany and more from Ireland and the colony distribution was changing – more Irish both in New York and Maryland. Of the American-born, the preponderance of migration was within the colony; of all those residing in Maine at recruitment, only 17% had been born outside the colony; of all those recruits born in Maine, only 10% were recruited elsewhere, and most movement within the colony was within counties (Table 3.1). The figures raise interesting historical questions. Why was immigration into South Carolina and New Hampshire so high? Why was there so much emigration from Pennsylvania? Why within Massachusetts did Hampshire and Berkshire attract so many immigrants, and why did Suffolk lose so many? The authors deduced that emigration was relatively high from the older, more densely populated coastal regions and particularly from cities, whereas the persistence rates were relatively high at the frontier – a pattern in contrast to the findings in the following century (Thernstrom, 1973) when the frontier regions showed lower persistence than cities. The authors generalize: 'The cities seemed to have been filling with foreigners and spewing forth both these foreigners and those born in the cities towards the frontier, to be replaced by new waves of immigrants from Europe.' Admittedly, many of the samples are small; they relate only to a section of the population, and so may not be entirely representative of the pattern of movement, but they are at least suggestive.

It is interesting to find such a pattern of movement that was essentially local in what was a new and developing society where many were still putting down roots. Certainly the strong tendency to remain in or near the birth locality is characteristic in longer established societies. This is well shown by comparison of data on two generations, namely the place of birth of an individual and the place of birth of his parents, which indicates the amount of movement in a generation, and such studies have been particularly useful for illustrating the conservatism of traditional rural

Table 3.1. *Percentage migration into and out of (a) the American colonies and (b) the counties of Massachusetts, as shown by enlistments in the late eighteenth century*

	Movement in	Movement out
(a) State		
Maine	17	10
New Hampshire	49	15
Massachusetts	15	22
Connecticut	14	21
New York	27	9
Pennsylvania	31	49
Delaware	31	15
Maryland	14	31
Virginia	13	28
North Carolina	34	12
South Carolina	78	10
(b) County		
Barnstable	15	8
Plymouth	6	17
Bristol	2	6
Norfolk	10	26
Suffolk	26	76
Essex	5	10
Middlesex	22	19
Worcester	30	14
Hampshire	42	8
Berkshire	50	37

communities. For example, such data from the Dalmatian island of Hvar show that 90% of the parents of those born in the inland villages had themselves been born in the same village as the offspring (Jovanovic *et al.*, 1984).

Migration is like Janus: it has two faces. There are those who emigrate and those who arrive, so that immigrants to one destination are not necessarily those who emigrated from elsewhere to do so. In these days of easy travel, the sheer difficulty and danger of actually moving long distances in times past is easily forgotten, quite apart from the perils on arrival and settlement. Not all the transatlantic immigrants survived the crossing. Those destined for Canada are well documented. For instance, the *Elizabeth and Sarah* sailed from Kilala in 1846 with 276 passengers, but 42 of them died during the voyage; the *Larch* sailing from Sligo in 1847 started with 440 passengers, of whom 108 died at sea, and 150 arrived with fever; the *Virginius* left Liverpool for Quebec with 476 passengers, of whom 158 died on the voyage and 106 were landed sick. Many ships brought fever; of the eight vessels that arrived on 21 May at Quebec there were no fewer than 430 fever cases amongst the passengers. Because of this all vessels were required on arrival to stay in quarantine; the *Agnes* which arrived in St Lawrence with 427 passengers had only 150 alive after a quarantine of 15 days. Of more than 100 000 emigrants who left the United Kingdom for Canada in 1847, by the end of the year no fewer than 20 000 had died in Canada, in addition to the 17 000 who perished during the voyage. On Grosse Isle there is a monument that states 'In this secluded spot lie the mortal remains of 5294 persons, who, flying from pestilence and famine in Ireland in the year 1847, found in America but a grave'. Many of these deaths were due to the poor hygiene, accommodation and overcrowding on the boats, and in particular to typhus which flourished in such surroundings.

But similar disasters on a smaller scale occurred amongst immigrants who were not subjected to such conditions. For example, the population of Pitcairn Island, deriving from the *Bounty* mutineers and Polynesians who founded it in 1790, grew rapidly in its earlier years, and particularly after the disruption of the initial period of unrest, strife, and murder. The island was completely isolated until 1808, and was very rarely visited until after 1825. But, by 1831 the islanders felt themselves restricted by the limited space, and the population of 87 moved to Tahiti. After a few months, however, having lost over a fifth of their numbers through disease contracted there, they returned to Pitcairn.

Demography of immigrant communities

Migrants are rarely demographically representative of the populations from which they derive, though the extent to which they vary depends on the nature of the migration. At one extreme are the groups entirely of younger adult males who migrate for labour, at the other are the whole communities, as when religious sects have moved. Rarely do groups of females, of children, or of old people migrate, and these are the subgroups usually in deficit in the age–sex pyramid, but little further generalization is possible.

Once migrants have settled in their new area, they may or may not retain their characteristic attitudes and cultural practices; the former is more likely when there is group migration and where there are settlement enclaves. An example is given by the Old Order Amish of Holmes County, Ohio (Cross & McKusick, 1970). This sect originated in 1693 in the canton of Berne, Switzerland, settled in Alsace-Lorraine and then, following severe persecution, began to migrate to the United States in about 1717. The first Amish to settle in Holmes County arrived in 1810. A survey census was made of the Holmes County community in 1964. They are characterized by two principal features: high standards of living and medical care, and prohibition of any form of birth control as a result of their strict religious beliefs. Consequently it is not surprising to find considerable fertility differences from the general population. The Amish crude birth rate for 1964 was 33.3 compared with 21.2 for the United States the same year, and the age-specific fertility rate was consistently high. The completed fertility averaged 7.19 pregnancies with standard deviation of 3.69, and the mean number of live births 6.33. The intrinsic rate of natural increase (r) was 0.30, indicating that the population was doubling in size every 23 years. It is not surprising that the age–sex pyramid with its broad base and narrow apex differed pronouncedly from that of rural non-Amish. Here then is an example of an immigrant community that has retained its fertility patterns over several generations.

While this is not unusual when emigrants colonize an empty territory, it is unusual when they enter an existing strong culture. Muthiah & Jones (1983) observed that in a number of overseas Indian communities during the 1950s and early 1960s, fertility was generally higher than in the states of their origin in India, but it declined more quickly during the following two decades than in the states of origin and is now lower. They suggested that traditional demographic transition theory is sufficient to explain this contrast between the fertility trends in India and the emigrants. Indian fertility remained high by comparison with the populations in the host countries, however, and particularly in Britain. This the authors attributed to the very low fertility in the indigenous population in the United Kingdom, and the recency of much Indian immigration, and hence lack of time for adjustment to the host fertility patterns.

Certainly there is such a process of adjustment. Another example of migration from one culture to another is that from the Pacific Polynesian island of Tokelau to New Zealand that began in the early 1960s, and was encouraged by the New Zealand government in the middle and later 1960s. The migrant population in New Zealand increased dramatically, from perhaps 30 persons in 1962, to some 1800 in 1972, and to about 3000 in 1982, including a substantial contribution by Tokelauans who migrated from Samoa, many of whom had never lived in Tokelau. Applying life table methods to survey supplemented with census and registration data, Pool *et al.* (1987) showed secular changes in the effect of migration on birth spacing and timing. In New Zealand the migrants married later and achieved their second birth at a later age; thereafter there was evidence of family limitation, there were differences in the rate of family building, and the more recent cohorts show increased tempo of earlier births and longer birth intervals. This change is in the direction of the patterns of family building of the New Zealand White population.

Similar adjustment seems to occur in rural-to-urban migrants. Lee & Faber (1984) recognized that the lower fertility among rural-to-urban migrants than among those who remain in rural areas may be due to three factors: selection of migrants, the disruptive effect of migration itself, and adaptation to urban constraints and norms. Using World Fertility Survey data from Korea sampled from the census enumeration districts, they enquired whether there was evidence of adaptation, after controlling for any selection by fertility preference. Fertility rates of rural–urban migrants before migration differed significantly from that of rural–rural migrants, but not of rural non-migrants when socioeconomic variables were controlled. Fertility rates of those migrating more recently to the town began to fall sooner after migration than among those who moved earlier. The decline persisted as the duration of urban residence increased. The study provides evidence that adaptation following rural–urban migration is a significant factor in the lower fertility of such migrants. Indeed, a three-stage model of adaptation of fertility to urban migration was proposed for Africa (Coulibaly & Pool, 1975). In the first stage, selective cyclical migration of the young and single may produce lower urban and rural fertility. In the second stage of family migration, urban fertility may well rise above rural, on account of the reduced cultural controls. The third stage may involve the adoption of host community reproductive norms and attitudes favouring contraception (Coulibaly & Pool, 1975).

Effect on the originating population

In the originating population there are both direct and indirect effects of migration on the level of reproduction. Direct effects are those due to a gain or loss in the number of females in the reproductive age groups caused by migration of females. Indirect effects are those influences exerted on the proportion of females who marry by the change in the sex ratio produced by a different level of net emigration among females and males. A smaller proportion of married females will lead to a lower total population fertility and hence a reduced level of reproduction.

Matthiessen (1972) examined the influence of emigration on reproduction in Denmark in 1840–1924. From life tables he calculated the expected number of females of reproductive age by five-year cohorts. Comparison of the expected with the observed number of females in each period demonstrated the relative change, positive in the cohorts 1840–1849, 1890–1894 and 1915–1919, but negative in all the remainder with a maximal loss of 6.5% in the cohort 1875–1879. The effect on the net reproduction rate was to reduce it from 1.226 to 1.146 in that cohort. Over the whole period the effect of emigration was an overall reduction of the net reproduction rate by 2.21%.

While in a large population such as Denmark the fertility effects of modest emigration are discernible but not massive, in a small population emigration may well be a critical factor in its extinction or survival. This is well illustrated by analysis of the effects of emigration on reproduction on the island of Tristan da Cunha in the South Atlantic. In Fig. 3.1, the plot of size of the population on 31 December each year, from its foundation in 1816 to 1975, shows two major episodes of emigration. In the first of these the numbers dropped from 103 at the end of 1855 to 33 in March 1857. This massive reduction seems to have been due to a combination of two principal factors: the death of one of the founder members, and the presence of a missionary. After the founder's death, the cohesion of the community appears to have relaxed, and 25 of his descendants left for

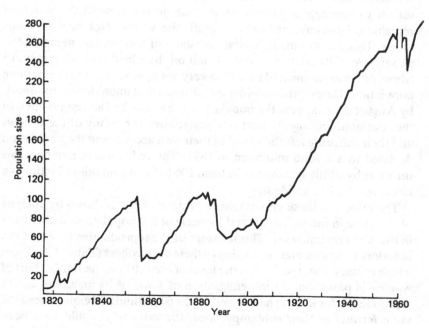

Fig. 3.1. The population size at 31 December each year.

America in 1856. The pastor who remained on the island until 1857 became increasingly convinced that emigration was necessary for the population. Whether this was true, or whether this was a projection of his discontent with his own lot there will never be known, but he noted that there were 'more than a dozen adult females here, with no prospect of a comfortable provision for life' and again, 'it will be a happy day when this little lonely spot is once more left to those who probably always were . . . its only fit inhabitants – the wild birds of the ocean'. Under his influence, when he departed another 45 islanders left with him for South Africa. The presence of a pastor of this opinion at a time when the population was reorganizing itself after the death of its dominant character can only be regarded as a combination of chance occurrences. The growth curve of the population size to that date shows no sign of flattening, and no indication that numbers were approaching the limit that the island could support.

The second bottleneck was neither quite so extreme nor quite so abrupt. It was triggered by a disaster. The island has no natural harbour, and any vessels that called stood offshore while the islanders put out to them in their small boats. Sometimes they also put out to board passing vessels for trade. On 28 November 1885, a boat manned by 15 adult males set off to intercept a passing vessel, but, in full view of the watching remaining islanders, vanished beneath the waves. Not one man was saved. This disaster made Tristan an island of widows and depleted the population of its adult providers. It left on the island four adult men, of whom one was insane and two were very aged, to support the remaining population. Despite the considerable distress that immediately followed, by August the next year the population as a whole had accommodated to the situation. During the next few years, however, many of the widows and their offspring left the island of their own accord, and the population declined to a second minimum in 1891. The reduction in numbers this time was by a little less than half, from 106 to 59. Again this reduction is a direct result of an accident.

The effects of these emigrations on reproduction is shown by analysis of variation in the rate of natural increase of the population, defined as *r* in the Volterra equation. This measure of the reproductive fitness of the islanders was calculated according to their birth cohort (Table 3.2). From a high initial value, the drop in the level of *r* to 0.013 in the third cohort of women is partly due to the emigration of some of its members before reproducing, for if these had stayed on the island and had experienced the same fertility as their contemporaries, the value of *r* would have been 0.033. It is in the fourth cohort (born 1840–1859) that the outlook for the population's future appeared black, for the rate of increase became

Table 3.2. *The intrinsic rate of natural increase in the Tristan population*

Cohort	Women	Men
x −1814	0.076	0.044
1815–1829	0.045	0.021
1830–1839	0.013	0.023
1840–1859	−0.006	−0.007
1860–1869	0.011	0.003
1870–1879	0.004	−0.002
1880–1889	0.010	−0.017
1890–1899	0.022	0.031
1900–1909	0.026	0.016
1910–1919	0.019	0.015

negative, and this was due primarily to the first large-scale emigration. Only five of the 26 females in this cohort stayed on the island to reproduce, and of these one left at the age of 21 after her second child was born. Indeed had the remainder not emigrated but stayed and experienced the same reproductive pattern as their companions, the value of r would have been +0.043 instead of negative. There was some recovery in the next cohort, but then again in the 1870–1879 group another serious decline to a negative value (−0.004), in which the succeeding cohort (1880–1889) also shares to some extent with the low positive value of 0.010. Again this decline is primarily due to the extensive second exodus. If this emigration had not occurred and the women had shared their contemporaries' fertility, the values of r for these two cohorts respectively would have been 0.031 and 0.037. Males show a similar pattern: the fourth cohort was particularly affected both by the first emigration (as boys) and by the boat disaster (as adults), the sixth and seventh cohorts were those most affected by the emigration following the boat disaster. These periods of negative rates show clearly the direct reproductive effect of the two relatively large emigrations from the island which despoiled the population of much of its reproductive potential. The slight differences in overall rate between the sexes is partly due to the fact that slightly fewer women than men contribute to the total offspring produced, since more women than men had more than one partner, and partly to the slightly later male age at reproduction.

Migration and genetics

The principal genetic relevance of migration is that it transfers genes from one locality to another. Migrants bring their genes into the territory inhabited by the recipient population, so that they are available for incorporation into the latter's gene pool if intermixture occurs. Intermixture, moreover, allows the transfer of genes from the host population to the gene pool of the migrants. There is thus simultaneous enhancement of the genetic variability within each of these populations and reduction of any genetic differences there may be between them; enhanced intrapopulation and reduced interpopulation variability. The way in which gene frequencies of such intermixing populations approach each other was calculated by Roberts & Hiorns (1962), applying a least-squares method of analysis, and other methods are those of Krieger *et al.* (1963) using a maximum likelihood solution, and Chakraborty (1975) based on the probability of gene identity. The migration process is thus the vehicle for one of the major mechanisms of evolution, defined as change in gene frequency of the population.

Immigration and the effects on the gene pool of the recipient population

An example of the genetic effects of immigration is provided by Crawford's classic study on the Black Caribs of Central America. This is based on genetic survey data and draws on early census and historical records for support. At the start of the sixteenth century the gene pool on St Vincent Island of the Lesser Antilles contained only Amerindian genes, derived partly from Carib and partly from Arawak peoples. From 1517 to 1646 an African component was added to the gene pool, deriving from African slaves, runaways or captives from nearby territories. After a series of vicissitudes, a number of the population, by now known as Black Caribs, were settled in British Honduras, Guatemala and Nicaragua. There followed rapid dispersal along one thousand kilometres along the coast of Central America into the present 54 distinct communities. There was subsequently some input of European genes, via the Creoles. Today, the Black Carib gene pool shows the undoubted presence of an African contribution, as shown for example by the presence of the haemoglobin S gene, the Fy^4 gene, the S^u gene of the MNS blood group system and the Gm haplotype $Gm^{z,a;b}$. The presence of the European contribution is shown by the Gm haplotype $Gm^{f;b}$, and the American Indian by the presence of the genes for the Diego blood group and albumin Mexico. Admixture estimates vary from one community of Black Caribs to another, but in general the Africans seem to have contributed between 60% and 80% of the gene pool, the Europeans up to 16%, and the American Indian 17% to 40%. The Black Caribs provide an example where the genes from immigrant populations predominate today over those from the original local population.

A remarkable feature of this population is their high degree of polymorphism at many loci, deriving from the heterogeneity of ancestry that migration brought. The number of distinct phenotypes in the Gm blood group system in a European population averages 10–12, whereas the Black Caribs of St Vincent show 42. The mean heterozygosity per locus of the Black Caribs (some 45%) considerably exceeds that in any other population yet studied, though strict comparison can only be made when the loci tested are the same in the different populations examined. But the increase in heterozygosity would be expected to enhance the hybrid vigour that may be expected in such a population, and it is possible that the high Darwinian fitness of the Black Caribs owes something to this as well as to their social organization.

A third feature that the Black Carib population demonstrates is the enhanced genetic adaptability that may emerge as a result of genes brought by migrants. There was heavy malarial infestation of this area until recently, primarily due to *Plasmodium vivax* and *Plasmodium falciparum*, carried by *Anopheles*. In the African gene pool there have evolved a number of genetic systems in which variants provide some protection against these malarial species. The advantageous systems include haemoglobin (genotypes AS and AC), and the Duffy blood group (homozygote FyFy). These genotypes were not present amongst the indigenous American Indian or the European ancestors of the population, but the genes were brought there by the African immigrants. These genes seem to have been a major factor in the evolutionary success of the Black Caribs and in their ability to colonize malarially infested regions. Here, then, is an example where the genes brought by migrants have altered the selective efficiency of the population in the area.

Effects of emigration on the gene pool of the population of origin

Genetic constitution

In a large population which has minimal individual emigration, the genetic effects will be negligible. They may be important or indeed critical in a smaller population. It is difficult to identify the effects of emigration from a population in terms of its gene frequencies. It would be necessary to know the gene frequencies in the population immediately before and after emigration occurred, requiring, respectively, foresight to know that it was about to take place, and hindsight to know that it had finished.

However, the effects can be identified if the genetic constitution of the population is specified not by the frequencies of particular alleles and genotypes, but by probable ancestral contributions, and this is possible when the full pedigree of the population is known. This procedure has been applied, for example, in the island of Tristan da Cunha in the South Atlantic, for which there are excellent records and where, as indicated in Fig. 3.1, there were two major episodes of emigration.

The effects of these emigrations on the genetic constitution of the population is shown in Fig. 3.2. Before the first emigration, 20 ancestors had contributed genes, their respective contributions to the gene pool varying from 0.005 to 0.137. The two greatest contributions were from the two original settlers (ancestors 1 and 2) and more than half the genes in the gene pool at the end of 1855 had been contributed by only five settlers. The effect of the first emigration is shown by comparison of the profile for 1857 with that for 1855. First, it deprived the population of all the genes from eight of its founder ancestors and a recent arrival, whose total contributions in 1855 had been more than one third, for genes from only 11 were to be found in the new gene pool. Secondly, there was a change in their relative contributions. Genes from two of the principal contributors (ancestors 6 and 7) completely disappeared. The greatest contribution was now from ancestor 4, followed by ancestors 3, 9 and 10. These four individuals together now contributed 60% of the genes in the new population compared with their previous total of less than 29%. The contributions of the first two settlers were halved, and that of a former minor contributor (8) multiplied nearly three-fold.

In the phase of increase that followed this first emigration, the contribution of one further ancestor was lost, but new contributions to the island's gene pool came from six more arrivals, so that by the end of 1884 the gene pool derived from 16 individuals. It was relatively little affected by the boat disaster itself, but the subsequent population size reduction had a

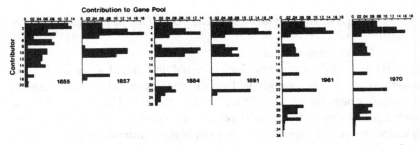

Fig. 3.2. The profiles of probable ancestral contributions to the poulation of Tristan da Cunha in selected years.

much more pronounced effect, shown by comparision of the profiles of 1891 with 1894. Again there was loss of all the genes from several contributors, and again there was rearrangement of the relative contributions of the remainder. All genes were lost from four relatively recent arrivals, so that the new gene pool derived from only 12 individuals. The greatest contribution was still that of ancestor 4, increasing from 0.14 to 0.19, and the second largest from her husband (ancestor 3) increasing from 0.09 to 0.14. But the third largest was from a much later immigrant woman (ancestor 22) which nearly doubled, while the former principal contributors 9 and 10 dropped from 0.14 to 0.08 and 0.13 to 0.06. In terms of genetic constitution, therefore, these emigrations (a) reduced the genetic variability of the population by eliminating all genes derived from an appreciable proportion of the ancestors, and (b) altered the actual and relative sizes of the contributions from other ancestors. Moreover, the effects of these emigrations persisted, as comparison with the profile of the 1961 figures shows. The general similarity of 1961 to 1891 implies that, apart from the contributions of the recent immigrants, the gene pool of the 1961 population derived its major features principally from the reduction in population that occurred between 1884 and 1891, and the modification that this brought then acted on the gene pool whose features chiefly derived from the effects of the earlier emigration.

Genetic structure

The effects of these emigrations can be traced not only on the genetic constitution, but also on the genetic structure of the population. The first emigration reduced the population size to such an extent that the choice of spouses, already restricted by the small population size, was reduced still further. The first consanguineous union occurred in 1854, and in this and in the second in 1856 there was little freedom of choice for the man concerned, respectively three and two unmarried women of marriageable age unrelated to him. But following the emigration, this choice disappeared, and by the time of the third consanguineous union in 1871, not only were there no non-relatives, there were indeed no other available women at all, and the same held at the time of the fourth consanguineous marriage in 1876. Thus in Fig. 3.3 the initial increase in the mean autosomal inbreeding coefficient is not attributable to the emigration, but the pronounced increase in 1872 and much of the general increase that follows is. Similarly, the increase in the sex-linked inbreeding coefficient of the early 1870s is so attributable. There is a similar increase in the mean autosomal coefficient with the second emigration, but this produced a diminution in the mean sex-linked inbreeding coefficient.

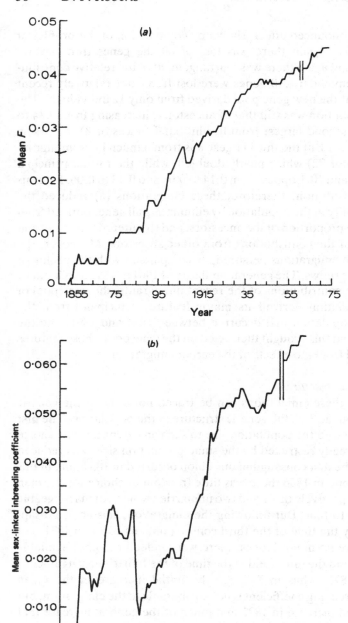

Fig. 3.3. (a) The evolution of the mean autosomal inbreeding coefficient. (b) The evolution of the mean sex-linked inbreeding coefficient.

The effect of the emigrations on the genetic structure is shown also by the analysis of kinship coefficients. The mean coefficient more than doubled with the first emigration (Fig. 3.4) and also increased, though to a lesser extent, with the second emigration.

This example is perhaps extreme, on account of the small size of the population and the magnitude of the emigration. However, for much of human history, sizes of breeding population were indeed small and the process here illustrated reflects a situation that must have occurred many times in the earlier phases of human evolution.

Types of migration and their genetic effects

Long-range migration

To assess the genetic relevance of migration it is important to distinguish migrant source, and particularly between long-range and short-range migration, though these clearly form a continuum. Long-range migration is usually more permanent, and in earlier days often entailed a complete cutting of ties with the ancestral population. No matter whether it represents movement of a few individuals continuing over a long period, or many individuals over a short period, the genetic effect is the same, to introduce new genes into the locality inhabited by the recipient population, and hence the opportunity of amalgamation of the gene pools. In view of the strong geographical gradients in gene frequencies, genes introduced by long-range migration are more likely to be at different frequencies from those in the recipient population and possibly entirely new to it. Massive movements for which there is documentation include the emigration from Europe for settlement in the Americas, of slaves

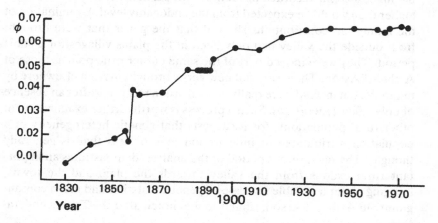

Fig. 3.4. The evolution of the mean kinship coefficient (ϕ).

from West Africa to the New World, of deportees from Britain to Australia, and as a result of these the gene frequencies in those continents today are very different from those in the indigenous population before the migrants' arrival. But most human migration is on a less heroic scale in terms of numbers and distance moved, and yet its genetic effect is considerable.

Short-range migration

Short-range movement consists mainly of the movement of individuals, usually associated with choice of spouse, marriage, and the couple's subsequent movement, which influences the local population into which their children are born. Such movement can be studied from the many records that exist, particularly in Europe, on birthplace of marriage partners, of parent and offspring, and place of death. Such movement circulates genes within the same or from very similar gene pools. An interesting demonstration comes from southern France.

Records of a small village and its surrounding populations in the French Pyrenees were analysed by a similar method to that applied in Tristan, examining the probable ancestral origin of genes. Serre, Jakobi & Babron (1985) calculated the new contributions each generation to the gene pool of Arthez d'Asson according to their geographical origin. They distinguished as a first category the genes of immigrant origin, brought by 1073 migrant ancestors from three villages on the plain outside the valley mouth (the great majority) and from two villages higher in the valley itself. Their second category related to genes already present in the genealogical founders of the population in the seventeenth century, of whom there were 1605. The results showed quite stable genetic isolation, the highest so far recorded in France, but, particularly important, much higher than would be expected from the endogamy level. Examination of the immigrant contributions showed that the genes that were brought from outside the valley had only been in the plains villages for a short period. They were indeed part of the same earlier gene pool as those of Arthez d'Asson. They were not new genes brought in from elsewhere by migrants, but instead were really of local origin returning after an absence of only a few generations. Such a process is worth further examination in other rural populations, for it suggests that genetic heterogeneity of a population attributable to immigration may be less than is generally thought. The most recent period of the analysis demonstrates an important rural exodus from the valley towards the plain and the towns, bringing about a notable increase of genetic isolation and mean consanguinity in Arthez d'Asson, similar to that noted after the Tristan emigrations.

The possibility of separating short- and long-range migration demo-graphically is suggested by the use of the migration density function. The distributions of distances between birthplaces of spouses, and of parents and offspring, tend to be L-shaped (gamma distributions), and to obtain the migration density function the frequency of migrants from a given distance is divided by the class interval of the distribution. The study of Hvar by Jovanovic *et al.* (1984) indicated a clear antimode to the distribution, interpreted as distinguishing between short- and long-range migration. Applying the function integrals of these two categories to the data for individual villages showed differences in the nature of the migration. For example, more females than males participated in short-range migration, more males than females in long-distance migration, as expected. The long-distance migration showed the influence of the specific geographical features and population densities of the wider region, and particularly the location and size of major cities on the mainland whence most long-range immigration derived.

Group migration

The Tristan example illustrates another important feature of the genetic effects of migration. The reduction of population in both episodes was initiated by accident, yet there was a large non-accidental component to the emigration. In both cases family groups emigrated, so that which genes and what proportions of them were lost from the population's gene pool were partly a random and partly a non-random array. The families that departed obviously included individuals who felt they could no longer accept the conditions on the island or its prospects for them, so the non-random component in part may be identified as selective. But the departure of family units rather than individuals brought about what may be termed a 'booster' effect, whereby although the genes lost from the population by accident were in actuality a random sample, the loss of some of them was exaggerated by consequent deliberate emigration of families.

Indeed, the nature of human society is such that often it is a group rather than one person who migrates. In the fission and fusion pattern of society so often seen at the tribal level, when a local group splits to form a new one, such records as there are show that close relatives often stay together and the split occurs along lineage boundaries (e.g. Neel & Salzano, 1967; Fix, 1975). This tendency applies both to long-distance migration and movement from one local population to another, and when it involves a group of relatives it has been termed kin-structure (Fix, 1978). Kin structure alters the effect that migration has on genetic differences between local groups. In general, migration of individuals

Table 3.3. *The half-life of*
convergence of subpopulations in
the absence of external migration

Population	Half-life (generations)
Bedik	2
!Kung Bushmen	2
Gainj	3
Papago	4
Bundi	4
Aland (all)	6
Gidra	18
Aland (pre-1900)	19
Makiritare	24
Bougainville	∞

takes a random sample of the gene pool. However, since members of a family are more likely to be similar in the genes they have, they are not a random sample, so when a group migrates the booster effect that it gives to the genes that are removed from the population distorts the random representation. Thus, lineal fissions can dramatically increase variation among local groups. It may well have been an important factor in the primaeval establishment of the patterns of human genetic variation seen today. However, despite the enhanced variation immediately following such a split, it too is subject to the homogenizing tendency of subsequent intergroup migration. The rate at which the variance among groups in several populations converges is shown in Table 3.3, based on the calculations by Rogers & Harpending (1986) of the half-life, the number of generations required to eliminate half of the variance introduced by initial fission. These results are an overestimate because the effect of external migration has been ignored; nevertheless, in the majority of instances the half-life is only a few generations. Using a computer simulation, Fix (1981) showed that group differences are reduced less by kin-structured migration than by unstructured migration. He documented the operation of this process amongst the Semai Senoi in Malaysia. But the effect of this process on genetic variation is still unknown, for in population genetic theory the migration that is considered is almost exclusively by a random sample of individuals, and rarely considers kin structuring.

Local migration and genetic structure

Recognition of the genetic importance of intrapopulation migration stems from Sewall Wright's earlier models (1940, 1943) of the genetic dynamics of population. These envisaged a major population as consisting of a number of smaller subpopulations, between which migration brought about gene flow, so that in each generation each subpopulation incorporated a small proportion of genes from others. Earlier Wahlund (1928) pointed out that a population is rarely a homogeneous social entity, and that within most there exist some boundaries to breeding, which may be geographical or social in nature. The concept of isolates and subpopulations within mainstream populations is important in population genetics. The size of the population, a critical variable in the evolutionary process, is considerably smaller than may appear at first sight. The large population postulated in elementary calculations of genetic equilibrium does not exist, but instead it is conceived of as comprising a series of smaller units. Where there is no migration between them, then in the absence of differential selection, mutation, and assuming random mating within each, this division alone would lead merely by chance processes to their genetic differentiation. This is brought about by the random variation in gene frequency that occurs from generation to generation, due to the sampling process of zygote formation, the extent of the sampling variation depending on the size of the population. Such divergence would accumulate, and lead to the loss or fixation of genes randomly in different subpopulations. The gene flow brought by migration reduces the rate of divergence, restricts the loss or fixation of genes, and so maintains variability within each subgroup. A third consequence of migration is reduction in the tendency to increased homozygosity within the population as a whole.

On these processes useful information is contributed by the analysis of migration as indicated by the data in the records. Comparison of the origin with the place of settlement of the migrants indicates the geographical distance over which they have moved, and the effect of geographical barriers and route ways on these movements. Secondly, comparison of the origins of spouses indicates the amount of spatial exogamy (marriage between people from different demographic units), and the distance over which marriages occur, while comparison of birthplaces of offspring with those of their parents shows the distance that their genes have travelled in a single generation. Such movement does not take place freely in all directions, and the records give some indication of its orientation. There are thus three components of movement, and hence of gene flow – the degree of exogamy, the distribution of marriage distances, and the direction of movement. These are of critical importance as determinants of the genetic structure of the population, controlling the extent of its genetic homogeneity or development of differentiated isolates, and the pattern of relatedness of its constituents, assessed both in terms of similarity of gene frequencies and in terms of the genealogical structure.

How limited and how local was the exogamy, and consequent gene flow until quite recent times is well shown in a study of a rural transect across mid-Northumberland (Dobson & Roberts, 1971; Dobson, 1973; Rawling, 1973; Roberts & Rawling, 1974) along the valley of the River Coquet. The number of marriages were examined in which both parties were from within the parish, one from outside, or both from outside, in the latter half of the seventeenth to the beginning of the nineteenth centuries, and it was found that never less than two-thirds of all marriages were contracted between partners who were both from within the parish. The figures range from 79–89% at the beginning of the period, to 68–75% at the end. This slight diminution was essentially due to a small increase in the number of unions in which the male was from outside the parish, and in these most of the movement occurred between adjacent parishes. For example, of the 1470 Rothbury marriage entries during the eighteenth century, 1168 (79%) were between two indigenous partners; in 172 (12%) the incoming partner was from an adjacent parish, so that less than 5% of marriage partners were drawn from anywhere more than a single parish width away from Rothbury. But, in fact, the localizing tendency was yet more extreme. Examination of the distribution of partners within the parish showed a striking preponderance of marriages in which both partners were resident in the same community within the parish (73% and 61% of intraparish marriages in Warkworth and Rothbury at the beginning of the century). This limited movement can be quantified by examination of the distance between birthplace of parent and of offspring. In Rothbury this averaged 6.1 km (s D 8.9), in Alwinton 9.2 km (s D 12.1) and in Felton 7.2 km (s D 10.9).

This marriage distance analysis suggests that there is a high level of genetic continuity within the parish, but some degree of isolation between parishes. The distribution of marriage distances is highly leptokurtic and, therefore, attempts to apply the uncorrected mean in estimates of effective neighbourhood size in Wright's isolation-by-distance models of population structure produce gross overestimates. Application of an island model to the three parishes of Alwinton, Rothbury and Felton, using data from the end of the eighteenth century, however, illustrates the effect of migration on rate of differentiation. Table 3.4 shows the total number of parents in each parish and the effective population size, taking

Table 3.4. *Effective population size and derived inbreeding estimates in mid-Northumberland parishes*

	Alwinton	Rothbury	Felton
Number of parents	233	540	410
Effective population size (N_e)	141	281	260
$\sigma^2_{dq}(q = 0.5)$	0.000887	0.000445	0.000481
$N_e m$	24.0	34.8	41.1
$\sigma^2_{dq}(q = 0.5)$ with migration	0.000611	0.000342	0.000341

into account the mean and variance of their number of offspring. The effective population size, N_e, in Alwinton is approximately 19% of the total population, a little less in Rothbury, and a little more in Felton. Certainly it is sufficient to prevent appreciable genetic drift of gene frequencies, σ^2_{pq}, even were there no marriage movement. For a gene of initial frequency 0.5, the chances are only one in 50 that the frequency would drift in a single generation by more than $\pm 0.069, 0.049,$ or 0.051 in the three parishes Alwinton, Rothbury and Felton respectively. When the migration that occurred between the parishes is also taken into account, $N_e m$, these figures reduce to $\pm 0.058, 0.043$ and 0.043. Appreciable differentiation of parish populations by drift would be still less likely.

The 'normalizing' importance of the genes brought by migrants is shown in Cartwright's (1973) study of Holy Island off the Northumberland coast. The present population of Holy Island can be divided into three categories. First, there are about a hundred individuals who were born on the island, who have many ties of kinship with other islanders, who have a characteristic island surname, and who have lived most of their life on the island. Secondly, there are the individuals who were born elsewhere but who have married islanders and have made their homes there so their offspring will be candidates for group I in the next generation. Thirdly, there are those who maintain holiday homes there to live in for periods of the year, retired people, and others who live on the island because of their occupation, and all these can be regarded as a transient population with no family ties to the island. From the parish registers similar categorization emerges throughout the population's history. The baptismal entries from 1578 to the present day were arranged into nuclear families and these families into cohorts. Each consisted of two distinct but overlapping populations. In one, the names of both marriage partners are known, one parent at least was born on the island so the name is reported in earlier cohorts, while the offspring in turn tended to remain on the island and their children recorded in later cohorts. The second population by contrast consists of individuals whose surnames are unfamiliar, have no traceable links with other families living on the island, the parents were not born there, and their offspring rarely remain on the island to become parents; in the few cases where they do remain it is only for one or two generations. The populations also show an occupational separation, the first being mainly farmers, fishermen or landowners and the second tending to be labourers, tradesmen or soldiers. These two contrasting populations represent, on the one hand, a stable and settled community and, on the other, a transient and poorer group. They are not entirely separate for the latter group often provides marriage partners for the other, and thus provides a buffer reservoir of incoming genes which may or may not be passed into the island's own gene pool.

Information on the sources of the new genes comes from parish marriage registers, the baptismal entries associated with each cohort and early census details. They show a large mean marriage distance (76.1 km) for all marriage entries up to the middle of the twentieth century, and 37.8 km at the turn of the nineteenth century. The incoming genes clearly are of varied origin. They arrive not because some of the island population travel great distances to look for marriage partners but more usually because new people arrive on the island from a variety of places and whose offspring or indeed themselves then marry in.

The genetic effect of this movement has obviously been appreciable. For example, in the ABO blood groups of the present population there are some 25% of group A_1 and 10% of A_2, frequencies normal for Northumberland. However, when the population is divided into the stable population, the new spouses and the transients, the stable population has only 14% A_1 and 11% A_2, a pronounced deficit of A_1. Those marrying in, however, have 50% A_1 and no A_2. It is the recent inflow of A_1 genes by those marrying in that is responsible for the normality of the frequencies in the total population.

Such large mean marriage movement distances, which have a large long-range proportion, are more characteristic of urban than of historical and rural populations, though even in modern cities many marriages are surprisingly local. The curve of marriage distance itself, and particularly its departures from a gamma distribution (on account of the irregularities of population distribution, and the overwhelming number occurring at zero distance) tends to say something of the nature of the population, and particularly when the additional element of direction of gene flow is included. In a recent genetic analysis (Roberts, Jorde & Mitchell, 1981), Cumbria was divided into six regions, and the migration patterns, as shown by comparison of birth places of parents and child, were examined. Table 3.5 shows the mean migration distance of mother and father by

Table 3.5. *Migration distance (km) of parents in Cumbria and direction of movement*

	Parent/offspring birthplace						
	Region						
	1	2	3	4	5	6	Total
Mean E–W movement							
Mother	+5.649	+6.081	−3.063	−3.777	+1.611	−0.953	+0.893
Father	5.498	7.327	−0.365	−3.217	0.086	−0.927	1.470
Mean N–S movement							
Mother	2.439	10.958	3.695	0.226	−1.722	−4.361	2.227
Father	0.784	9.638	3.177	−0.240	−1.829	−5.918	1.192
Mean migration distance (mother)	13.826	18.813	10.863	6.938	12.776	11.674	12.075
Variance	229.15	361.32	151.51	153.37	142.01	202.44	236.81
Mean migration distance (father)	13.995	18.478	9.717	6.705	5.886	12.803	11.887
Variance	222.55	309.79	139.57	118.52	68.08	247.22	220.34

Key of regions: 1, Eden valley; 2, Carlisle city; 3, the northern plain; 4, west coast industrial area; 5, central Lake District; 6, southern periphery.
+ = west-to-east, south-to-north.
− = east-to-west, north-to-south.

Table 3.6. *Kinship between Cumbrian regions as calculated from birthplaces of parents and children*

		Region					
	Region	1	2	3	4	5	6
Estimated from:							
Father/	1	729	598	590	433	526	420
child	2		664	761	490	551	411
	3			1045	537	564	402
	4				625	508	374
	5					816	473
	6						493
Mother/	1	955	839	835	679	783	634
child	2		885	978	723	798	611
	3			1293	775	865	597
	4				880	769	576
	5					879	615
	6						708

Values given are kinship coefficients $\times 10^{-7}$.
See Table 3.5 for key to regions.

region. Carlisle city (region 2) has the highest mean migration distance for both sexes and also the highest variance, as expected for an urban centre; region 4 (the west coast industrial area) shows low mean migration distances for both sexes, reflecting the conservatism of these mining and industrial area families, and region 5 (the central Lake District) the lowest paternal mean and the lowest variances for both sexes, not unexpected in a traditional farming area. If such migration data are assumed to be constant from one generation to the next, an equilibrium situation with population size is reached and kinship coefficients to specify the relationships between pairs of regional populations can be calculated from the migration rates between them, the effective population size of each, and long-range immigration (Table 3.6). Region 6, the southern periphery, shows the lowest kinship with each of the other regions, suggesting the existence of some disincentive to migration between region 6 and the others; its links, such as there are, lie with region 5 (the central Lake District). Region 4 is the second most remote from the majority of the others, again suggesting the existence of a barrier, and its links lie especially with the adjacent regions 3 (the northern plain) and 5. The directions of gene flow, calculated by rectangular coordinates, show that latitudinal gene movement in regions 1 (the Eden valley), 2 and 5 was from west to east, and in regions 3, 4 and 6 from east to west. Longitudinal movement was north to south in regions 5 and 6, and south to north in 1, 2 and 3. The distance of gene flow suggested by the non-directional migration data is in fact an overestimate, movement in one direction being largely cancelled by that in the opposite. That due respectively to maternal and paternal movement averages 2.4 and 1.9 km respectively.

Conclusion

This essay shows how records obtained for one purpose, usually administrative, can be applied for quite another, to help understand the biology of migration. Such records can show very little regarding the biological qualities of migrants, and though it is well recognized that they differ from the sedentes left at home and from the recipient populations whom they join, it is by no means clear how far the differences are due to the effects of their new environment, and how far to their intrinsic differences. Nor can the records indicate very much of the effects of the migration process on the immigrants themselves; it may produce a greater sense of personal and political freedom, or it may be comparable to a process of bereavement – for some, a profound disturbance from which they never recover. To answer this type of question, specific surveys are required. The records do, however, illuminate some features of the migration process, particularly some demographic and genetic features that have tended to receive relatively little attention, and it is these that have been mainly considered here.

Migration is never absent, even in the most secluded of populations, but the amount until quite recently in settled communities was relatively small. Apart from the most local movement, the process carried appreciable risk. Once established, the immigrant communities may retain their own demographic patterns, but more usually adapt to those of the host population. But the demographic effects, both on the host population and the parental population from which the migrants derive, depend very much on the amount of migration. Movement that would be quite unnoticed in a larger population may have massive effects on a small one.

Migration is of considerable genetic relevance. It is the vehicle for the mechanism of evolution that today is producing the greatest evolutionary effect, allowing the incorporation of new genes into established gene pools, enhancing intrapopulation and reducing interpopulation variability. The new genes brought in may come to predominate proportionally over those of the host population, but the biological efficiency of the new hybrid population may well be enhanced. Certainly appreciable emigra-

68   D. F. Roberts

tion from a small population may have considerable effect, both on its genetic constitution, reducing the genetic variability, and on the genetic structure, reducing the genetic variability. Different types of migration – short-range, long-range, kin-structured – have different genetic effects. The genetic heterogeneity due to short-range migration may be less than at first appears, on account of the cyclical nature of movement, and because movement in one direction may well cancel out that in the opposite.

The examples given illustrate some new and little-used ways in which recorded data may be used that appear to deserve wider application.

References

Baines, D. E. (1972). The use of published census data in migration studies. In *Nineteenth-Century Society*, ed. E. A. Wrigley. Cambridge: Cambridge University Press.
Cartwright, R. A. (1973). The structure of populations living on Holy Island, Northumberland. In *Genetic Variation in Britain*, ed. D. F. Roberts & E. Sunderland pp. 95–107. London: Taylor and Francis.
Chakraborty, R. (1975). Estimation of race admixture. *American Journal of Physical Anthropology*, **42**, 507–12.
Coulibaly, S. & Pool, I. (1975). Un essai d'explication des variations de la fecondite en Haute-Volta et Ghana. *Population et Famille*, **34**, 1.
Crawford, M. H. (1984). *Black Caribs: Case Study in Biocultural Adaptation*. New York: Plenum.
Cross, H. E. & McKusick, V. A. (1970). Amish demography. *Social Biology*, **17**, 83–101.
Devis, T. L. F. & Southworth, N. R. (1984). The study of internal migration in Great Britain. In *Migration and Mobility*, ed. A. J. Boyce, pp. 275–300. London: Taylor and Francis.
Dobson, T. (1973). Historical population structure in Northumberland. In *Genetic Variation in Britain*, ed. D. F. Roberts & E. Sunderland, pp. 67–82. London: Taylor and Francis.
Dobson, T. & Roberts, D. F. (1971). Historical population movement and gene flow in Northumberland parishes. *Journal of Biosocial Science*, **3**, 193–208.
Fix, A. G. (1975). Fission–fusion and lineal effect: aspects of the population structure of the Semai Senoi of Malaysia. *American Journal of Physical Anthropology*, **43**, 295–302.
Fix, A. G. (1978). The role of kin-structured migration in genetic micro-differentiation. *Annals of Human Genetics*, **41**, 329–39.
Fix, A. G. (1981). Kin-structured migration and the rate of advance of an advantageous gene. *American Journal of Physical Anthropology*, **55**, 433–42.
Grundy, E. (1986). Migration and fertility behaviour in England and Wales. *Journal of Biosocial Science* **18**, 403–23.
Jovanovic, V., Macarol, B., Roberts, D. F. & Rudan, P. (1984). Migration on the

island of Hvar. In *Migration and Mobility*, ed. A. J. Boyce, pp. 143–60. London: Taylor and Francis.

Koertvelyessy, T. (1983). Demography and evolution in an immigrant ethnic community. *Journal of Biosocial Science*, **15**, 223–36.

Krieger, H., Morton, N. E., Mi, M. P., Azevedo, E., Freire-Maia, A. & Yasuda, N. (1963). Racial admixture in northeastern Brazil. *Annals of Human Genetics*, **29**, 113–25.

Lee, B. S. & Faber, S. C. (1984). Fertility adaptation by rural–urban migrants in developing countries. *Population Studies*, **38**, 141–55.

McKusick, V. A. (1978). *Medical Genetic Studies of the Amish*. Baltimore: Johns Hopkins University Press.

Matthiessen, P. C. (1972). Replacement for generations of Danish females – 1840/1844 – 1920/1924. In *Population and Social Change*, ed. D. V. Glass & R. Revelle. London: Arnold.

Muthiah, A. & Jones, S. W. (1983). Fertility trends among overseas Indian populations. *Population Studies*, **37**, 273–99.

Neel, J. V. & Salzano, F. M. (1967). Further studies on the Xavante Indians. *American Journal of Human Genetics*, **19**, 555–73.

Pool, I., Sceats, J. E., Hooper, A., Huntsman, J., Plummer, E. & Prior, I. (1987). Social change, migration and pregnancy intervals. *Journal of Biosocial Science*, **19**, 1–15.

Rawling, C. P. (1973). A study of isonymy. In *Genetic Variation in Britain*, ed. D. F. Roberts & E. Sunderland, pp. 83–93. London: Taylor and Francis.

Roberts, D. F. & Hiorns, R. (1962). The dynamics of racial admixture. *American Journal of Human Genetics*, **14**, 261–77.

Roberts, D. F., Jorde, L. B. & Mitchell, R. J. (1981). Genetic structure in Cumbria. *Journal of Biosocial Science*, **13**, 317–36.

Roberts, D. F. & Rawling, C. P. (1974). Secular trends in genetic structure: an isonymic analysis of Northumberland parish records. *Annals of Human Biology*, **1**, 393–410.

Rogers, A. R. & Harpending, H. (1986). Migration and genetic drift in human populations. *Evolution,* **40**, 1312–27.

Serre, J. L., Jakobi, L. & Babron, M. C. (1985). A genetic isolate in the French Pyrenees: probabilities of origin of genes and inbreeding. *Journal of Biosocial Science*, **17**, 405–14.

Thernstrom, S. (1973). *The Other Bostonians*. Cambridge, Mass.: Harvard University Press.

Villaflor, G. C. & Sokoloff, K. L. (1982). Migration in Colonial America. *Social Science History*, **6**, 539–70.

Wahlund, G. (1928). Zusammensetzung von Populationen und Korrelations erscheinungen von Standpunkt der Vererbungslehre aus betrachtet. *Hereditas*, **11**, 65–106.

Wright, S. (1940). Breeding structure of populations in relation to speciation. *American Naturalist*, **74**, 232–48.

Wright, S. (1943). Isolation by distance. *Genetics*, **28**, 114–38.

4 Models of human migration: an inter-island example

P. D. RASPE

As explained elsewhere in this volume (Chapter 6) models of migration fall into two general classes: distance models and island models. Human beings are distributed over the surface of the globe in patterns that are neither homogeneous nor completely aggregated. Clustering is observed at every level; in houses, in towns, in cities and conurbations. Although neither type of model exactly fits the human situation, users of both types of model believe that the approximations are good enough to give a true picture of the kinds of differences existing between human populations and of differences in migration rates or distances when compared to those incorporated in the models. For this reason, distance models have sometimes been applied to nucleated as well as evenly distributed populations and island models have been applied to regions of more or less continuous population distribution as well as to regions with discrete non-overlapping demes.

In this chapter I shall describe some properties of island models and apply these models to a group of islands, five of which are at present inhabited, thus comprising five discrete populations separated from each other by water. The findings from the application of such models will then be considered in the light of other studies to which island models have been applied, even though the assumptions of discrete 'islands' are imperfectly met.

Exchange of individuals and movement between populations is one of the important determinants of population structure. Whenever the frequencies of various alleles at a locus differ in a group of migrants from those of the host population, changes in the genetic composition of those populations will result. Such movement is only of genetic consequence when the migrating individual contributes his or her genes to subsequent generations of the host population. Therefore, the only movement of real genetic consequence is that associated with marriage and subsequent child bearing. A number of models have been developed to determine the genetic outcome of particular patterns of marital movement which make use of the actual behavioural data. These models are often classified according to whether they take into account random effects due to finite population size. Models which do not include random elements are said to be 'deterministic', assuming infinite population size, so that the frequency of a gene at the time of conception is equal to the probability of encountering that gene in the previous generation. Models in which finite population size is assumed, and in which the frequency of genes within a generation is a sample of the genes of the parental generation, are called 'stochastic' models. In both types of model the marital exchanges are compounded in the form of a migration matrix and their effects accumulated by multiplying repeatedly against a matrix of hypothetical gene frequencies or ancestor frequencies.

Such models are of two main subtypes. The models of Bodmer & Cavalli-Sforza (1968), and Imaizumi, Morton & Harris (1970) simulate a group of populations diverging from one another under the stochastic forces of drift, but whose divergence is limited by migration between the populations. Alternatively, the model of Hiorns *et al.* (1969) envisages a group of populations moving from maximum heterogeneity to maximum homogeneity under the deterministic effect of marital exchange, and demonstrates the pattern by which populations come to share common ancestry.

The model of Bodmer & Cavalli-Sforza (1968) uses a migration matrix in conjunction with estimates of effective populations size to predict the amount of spatial gene frequency variation in terms of the covariance between the populations in the system. An analogous approach has been taken by Imaizumi *et al.* to investigate the effects of drift and migration in terms of the coefficient of kinship, defined by Malécot (1948) as the probability that two allelic genes selected randomly from each of two populations are identical by descent. Under the model of Hiorns *et al.*, the effects of drift and selection, which may be significant in small populations, are ignored, since the concern of this approach is to show how many generations it will take, given a particular pattern of marital exchanges, for each pair of populations in the group to become homogeneous in terms of the distribution of common ancestry.

A model of each subtype can be used to investigate the consequences of the observed marital migration between populations. The deterministic model of Hiorns *et al.* is used to investigate the patterns of increasing homogeneity within the population. These patterns help us investigate the concept of genetic relatedness between populations in terms of increasing common ancestry resulting from genetic exchange. The model assumes a discrete series of populations which have no shared ancestry, but between which there is now gene flow in the form of mate exchange. Calculations are made of the number of generations required, under this particular .pattern of mate exchange, for the population to become homogeneous in terms of sharing a similar distribution of common ancestors, and the pattern by which such relatedness develops is illustrated. This is achieved by accumulating the effects of the observed mate exchanges through iterative powering of the matrix. For practical purposes homogeneity is defined as having 95% common distribution of ancestor frequencies.

The alternative model of Imaizumi *et al.* has been used to compare the overall effects of marital movement when genetic drift is taken into account. They have demonstrated that the variances and covariances in gene frequencies between groups can be expressed in terms of the coefficient of kinship. The recurrence relation demonstrated by Malécot (1950) is derived; this allows the calculation of kinship coefficients within and between populations on the basis of the probabilities of randomly selected genes being identical by descent, and the probability of a random gene substitution from the equilibrium gene pool because of long range migration. This is iterated over time in order to examine the approach to equilibrium values. This model has advantages over the more widely used model of Bodmer & Cavalli-Sforza because the formulation in terms of kinship enables the calculation of a measure of genetic distance proposed by Morton *et al.* (1971) – the coefficient of hybridity – which was calculated for each pair of populations, using the coefficients of kinship after 50 generations derived from the above model. This parameter indicates the extent to which the combined population deviates from panmixia. Therefore, the larger the value, the greater the difference between the two populations.

These models are formulated on the basis of a number of assumptions which may not be expected to pertain in the real world: that migration patterns remain constant through time; that migrants are randomly selected from the populations from which they derive, i.e. that they are genetically representative of their populations of origin; and, as noted above, that the populations concerned are permanent, clearly demarcated entities.

The Isles of Scilly

The Isles of Scilly consist of a group of islands, islets and rocks situated approximately 38 km west-south-west of Land's End, Cornwall (Fig. 4.1). The five largest islands – St Mary's, Tresco, St Martin's, St Agnes and Bryher – are at present inhabited, and a sixth island, Samson, supported a small population until 1855 (Bowley, 1968). These few

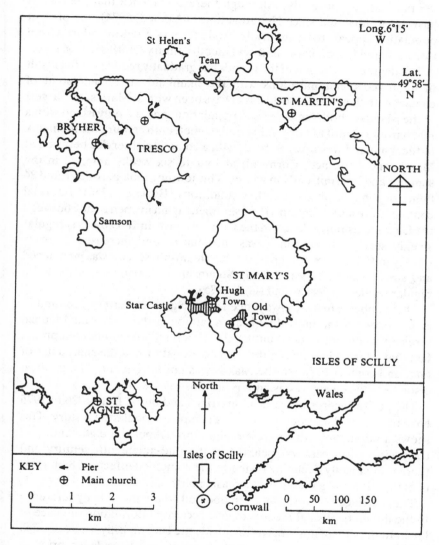

Fig. 4.1. Map of the Isles of Scilly.

individuals have been included with the population of nearby Bryher for the purposes of this study. Although there is evidence that the islands have been inhabited since Neolithic times (Ashbee, 1974), the present population appears to be largely descended from settlers introduced from Cornwall and Devon in the mid-seventeeth century (Borlase, 1756).

The degree of isolation of the islands has gradually reduced as the result of improved communications with the mainland. The main point of contact with the outside world has always been with St Mary's, the largest of the islands in both land area and population, and the only island with a real harbour. Until the beginning of the nineteenth century the main link to the mainland was by open boat, with a commercial service between St Mary's and Penzance, Cornwall, at four to six weeks' interval in the summer and intermittently in winter. This journey took between 6 and 24 hours depending on the weather conditions (Borlase, 1756). Informal contacts were also maintained by fishermen, again in open boats, between the Isles and fishing villages on the Cornish coast. In 1803 the first regular weekly service to Penzance was introduced, and increased to twice weekly in 1848. In 1858 a thrice-weekly steamship service was introduced and since 1920 the Isles of Scilly Steamship Company has provided a regular service all year round (Gill, 1975).

There appears to have always been a much greater degree of communication between the individual islands, although the inter-island launch service was not introduced until 1923. Heath (1750) commented on the fact that all island families either owned a boat or had the shared use of one, and contact between the islands was not infrequent. The greatest distance between any two islands is only just over 8 km.

The parish records of baptism, marriage and burial from 1726 to 1975 have been linked in order to provide a complete population history. This shows a population which grew rapidly throughout the eighteenth and early nineteenth century, peaking in the mid-nineteenth century and thereafter steadily declining owing to the combined effects of high rates of emigration and a gradually decreasing birth rate.

The marriage records have been updated where possible by reference to the linked baptismal records to give accurate information on place of birth. In the case of migrants from outside the Isles of Scilly where no such linking was possible, the information on place of residence prior to marriage was retained as an indicator of place of birth. Marriages prior to 1750 could not be linked to baptism records which only commenced in 1726. This yielded a file of 1361 marriage records giving the birthplaces of both husband and wife, from which marital exchange matrices were derived. Residence after marriage in the Isles of Scilly has been predominantly patrilocal and therefore the matrices of husband–wife birthplaces are considered to give an accurate indication of genetic exchange. In order to investigate the effects of the changing demographic patterns over time, migration matrices were derived for each of the time periods 1750–1799, 1800–1849, 1850–1899 and 1900–1975. The migration matrix for 1900–1975 contains the smallest number of marriages and is highly biased towards those who married within the Isles of Scilly, and particularly within the same island. This is primarily due to the exclusion of a high proportion of marriages which presumably involved a migrant from the outside world, but for whom no information on birthplace or place of residence prior to marriage could be obtained. For this reason the 1900–1975 period has not been included here. In addition, a total migration matrix was obtained by combining the matrices for the three time periods from 1750 to 1899.

The matrices given in Table 4.1 give the birthplace of the wife in the rows and the birthplace of the husband in the columns. Since residence after marriage is predominantly patrilocal, this can be considered to represent migration from the rows to the columns. Migration from outside the Isles of Scilly has been divided into migrants from Cornwall and Devon, who tended to be twice as numerous as the others, and those from elsewhere. Because of the ancestral history of the population, migrants from the adjacent counties of Cornwall and Devon are likely to be more similar genetically to the islanders than migrants coming from elsewhere.

The magnitude of migration is most concisely measured by the extent of exogamy, or conversely, endogamy in a population. When the Isles of Scilly are considered as a single population, the percentages of endogamous marriages are 97%, 92%, and 78% for the time periods 1750–1799, 1800–1849, and 1850–1899 respectively. This indicates a high degree of isolation from the outside world. However, the migration matrices include only those marriages where the birthplaces of both husband and wife are known, and a large proportion of marriages thereby excluded might be expected to involve a migrant. Such figures should therefore be considered as upper limits of endogamy, rather than accurate estimates.

The frequencies of endogamous marriages for each island by time period are given in Table 4.2. These are calculated with respect to the population of each island by dividing the number of persons who married endogamously within an island by the total number of persons born on that island. The small number of people born on Samson are included with the nearby Bryher population, so that the few marriages between these adjacent islands are treated as endogamous.

Table 4.1. *Migration matrices for the Isles of Scilly by time periods*

	1750–1799							1800–1849							1850–1899							1750–1899						
	SM	TR	SA	MT	BY	CD	EW	SM	TR	SA	MT	BY	CD	EW	SM	TR	SA	MT	BY	CD	EW	SM	TR	SA	MT	BY	CD	EW
SM	96	4	8	5	1	1	3	191	17	10	5	5	7	2	108	17	19	10	9	13	5	385	38	37	20	15	21	10
TR	3	34	0	4	3	0	0	27	94	6	1	5	5	0	19	20	2	4	5	2	4	49	148	8	9	13	7	4
SA	9	7	31	4	3	0	0	12	5	42	2	12	0	1	9	2	16	5	4	4	2	30	14	89	11	19	4	3
MT	5	5	2	20	3	0	0	9	5	4	42	1	0	0	8	2	1	17	1	2	0	22	12	7	79	5	2	0
BY	2	9	0	1	10	0	0	5	7	1	1	20	1	1	7	2	3	2	11	2	1	14	18	4	4	41	3	1
CD	1	1	0	0	0	1	0	16	0	0	0	2	5	0	19	4	1	2	2	3	3	36	4	2	4	2	8	3
EW	0	0	0	0	0	1	0	1	1	3	0	0	0	2	10	1	1	3	1	1	1	11	2	1	3	1	2	3

Wife's birthplace in rows and husband's birthplace in columns:
Abbreviations in this and subsequent tables: SM, St Mary's; TR, Tresco; SA, St Agnes; MT, St Martin's; BY, Bryher; CD, Cornwall and Devon; EW, Elsewhere.

Table 4.2. *The frequency of endogamous marriages
for each island by time periods*

	1750–1799		1800–1849		1850–1899	
	n	freq	*n*	freq	*n*	freq
SM	235	0.817	495	0.772	361	0.598
TR	104	0.654	265	0.709	104	0.385
SA	96	0.646	140	0.600	85	0.376
MT	69	0.580	114	0.737	74	0.459
BY	42	0.476	68	0.500	60	0.367
Total	546	0.700	1082	0.713	648	0.503

See Table 4.1 for abbreviations.

The frequency of island endogamy tends to be fairly high in all time periods, indicating a considerable degree of isolation for each of the islands. There is a marked decrease in the frequency of endogamy throughout the nineteenth century. There also appears to be an association between the island population sizes and the frequency of endogamy, in that the larger populations have the higher proportions of endogamous marriages, indicating that those from smaller populations are more frequently forced to look for spouses from elsewhere. The main exception to this pattern is the island of St Martin's which had the second smallest population in the nineteenth century but the second highest proportion of endogamous marriages. The correlation between population size and proportion of endogamous marriages is significant at the 5% level only for the period 1750–1799.

The frequency of inter-island marriage is given in Table 4.3 for all pairs of islands, together with the inter-island distances by sea. These are calculated with respect to the total population, by dividing the number of persons who married between a pair of islands by the total population size. Again the few individuals born on Samson are included with Bryher. Information on birthplace is available by island only, so that distances by land, which could be up to 3 km on St Mary's, could not be analysed. However, as travel by sea was undoubtedly more arduous, average distance by sea between islands is used as a measure of the ease of access between subpopulations.

There does not appear to be any consistent pattern of association between pairs of populations over time. For example, St Mary's and Tresco have low rates of exchange in 1750–1799, but the highest rates for subsequent periods. St Martin's and Bryher have the lowest rates in 1750–1799 but the second highest rates in 1850–1899. There is a slight

Table 4.3. *The frequency of inter-island marriage for all pairs of islands for each time period*

	Distance (km)	1750–1799	1800–1849	1850–1899
SM–TR	4.1	0.025	0.078	0.092
SM–SA	2.8	0.061	0.039	0.071
SM–MT	5.3	0.036	0.043	0.046
SM–BY	4.8	0.011	0.012	0.041
TR–SA	4.9	0.025	0.019	0.010
TR–MT	3.4	0.033	0.011	0.015
TR–BY	0.7	0.043	0.018	0.018
SA–MT	8.1	0.022	0.011	0.015
SA–BY	6.5	0.021	0.023	0.018
MT–BY	6.8	0.014	0.004	0.077

See Table 4.1 for abbreviations.

tendency for the overall amount of marital exchange to increase throughout the nineteenth century. Only the Tresco–St Agnes, Tresco–Bryher and St Agnes–Bryher rates decrease over this period. There appears to be a weak association between exchange rates and the combined population sizes, but because all the rates are low, small fluctuations in the number of marriages can disturb any association. The correlations are only significant at the 5% level for the period 1800–1849. There is the suggestion of a further weak association between geographic distance and the frequency of exchange. The correlation between frequency of exchange and the inverse square of the distance is significant at the 5% level only for the earliest period, 1750–1799. Partial correlations indicate that while in the eighteenth century the effect of distance is the strongest cause of the observed association, this is reversed in the nineteenth century when the effect of combined population size is primarily responsible for the associations.

The rates of endogamy are high, but overall endogamy is within the range of rates reported for comparable British island populations, such as Sanday in the Orkneys (Boyce, Holdsworth & Brothwell, 1972), and Lewis and Harris (Clegg, 1975) and Barra (Morton *et al.*, 1976) in the Outer Hebrides. The inter-island exchange rates are highly variable, showing no consistent trends over time, but a general trend towards an increase in overall inter-island movement over time is apparent. Such progressive breakdown in isolation has been documented for a variety of populations. Modern exceptions to this pattern are rare and are possibly exemplified by northern Sweden where a process of increasing isolation has been reported (Beckman, 1980).

It appears, therefore, that although inter-island distances may have represented barriers to mating in the eighteenth century, for most of the period under investigation the distances between islands did not impose any serious constraint on inter-island movement, and the entire population of the islands can be considered as forming the mate pool.

The deterministic model

The results of the analysis by the method of Hiorns *et al.* (1969) are shown in Table 4.4 for the time periods 1750–1799, 1800–1849, 1850–1899 and the total population for 1750–1899. When the twentieth century data were included with the total population the results were not substantially different from those for the combined populations 1750–1899 which are given here.

There is a marked decrease in the number of generations required for all the islands to share 95% common ancestry under the model. In 1750–1799 this would take 28 generations, in 1800–1849 it would take 19 generations and in 1850–1899 only 11 generations. This reflects the reduction in isolation of the Isles over time and the steady increase in marital migration between islands and with Cornwall and Devon and the outside world. The same trend was observed in the endogamy and exogamy rates described above. When the total population for the eighteenth and nineteenth centuries is considered, 14 generations elapse before 95% homogeneity is reached. When the effect of migration from the outside world is excluded from the analysis, that is, when the five islands are considered as a closed system, the length of time to attain 95% homogeneity is only very slightly increased, to 29, 19 and 12 generations respectively for the three time periods, and to 15 generations for the total population. This would indicate that the increasing homogeneity of the population over time is primarily due to the increase in inter-island marital movement, and that the effect of immigration from the outside world is negligible.

Fig. 4.2 shows the way in which homogeneity develops. The proportion of common ancestors averaged over all pairs of populations are plotted against generation until equilibrium values are reached. In 1750–1799 the development is fairly slow with 50% homogeneity attained after 8 generations while in 1800–1849 this level is passed after only 6 generations. In 1850–1899, 50% homogeneity is attained in less than 3 generations. The dotted line indicates the development of average homogeneity from the data for all marriages from 1750 to 1899. The curve falls between those for

Table 4.4. Numbers of generations required to reach 95% homogeneity between all pairs of islands by time period

1750–1799							1800–1849							1850–1899							1750–1899						
SM	TR	SA	MT	BY	CD	EW	SM	TR	SA	MT	BY	CD	EW	SM	TR	SA	MT	BY	CD	EW	SM	TR	SA	MT	BY	CD	EW
28							15							8							13						
22	21						17	16						9	9						13	12					
24	18	14					18	18	12					10	10	18					14	14	13				
28	9	21	18				16	13	18	17				9	10	8	10				13	10	11	13			
22	22	10	14	22			9	16	18	17	17			6	8	8	10	9			7	13	13	14	13		
21	22	10	15	22	9		17	13	9	17	13	18		17	16	9	17	13	18		8	12	12	13	13	7	

See Table 4.1 for abbreviations.

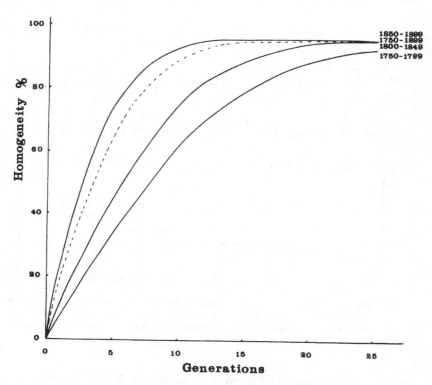

Fig. 4.2. Degree of homogeneity achieved (%). Unbroken line, rate during 50-year period; broken line, rate during the 150-year span.

the first and second halves of the nineteenth century. In this case 50% homogeneity is reached in less than 4 generations.

Table 4.4 shows the pattern of increasing relatedness between all pairs of populations, expressed in terms of the number of generations before each pair of populations share 95% common ancestry. This same information is shown graphically in Fig. 4.3, where the minimum number of generations required under the model for each population to become homogeneous with neighbouring populations is expressed in the form of a dendrogram. No consistent patterns of relatedness over time are apparent. In 1750–1799, Bryher and Tresco, the most closely associated islands geographically, are the first to become 95% homogeneous in terms of shared ancestry, but in 1800–1849 Bryher is more closely associated with St Agnes. In 1750–1799 St Agnes clusters with Cornwall and Devon and the outside world while St Mary's becomes 95% homogeneous with them after 21 generations. In 1800–1849 St Mary's

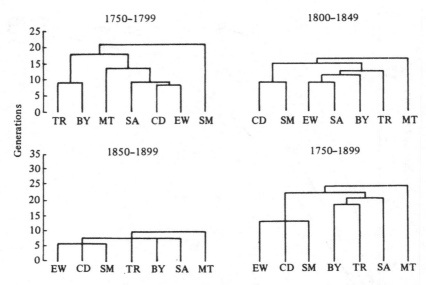

Fig. 4.3. Patterns by which relatedness (95% homogeneity) between islands develops as a result of inter-island marriage. Abbreviations: TR, Tresco; BY, Bryher; MT, St Martin's; SA, St Agnes; CD, Cornwall and Devon; SM, St Mary's; EW, elsewhere.

merges with Cornwall and Devon at the same time that St Agnes merges with the outside world, after only 9 generations. By 1850–1899 St Mary's is most closely associated with Cornwall and Devon and the outside world, sharing 95% common ancestry with them after only 6 generations. While in subsequent periods St Martin's is the last population to share common ancestry with all others, in 1750–1799 this place is occupied by St Mary's.

The main reason for this lack of consistency in the patterns of increasing relatedness of the population over time is the fairly small population size and therefore the magnified effect of even a small number of marital exchanges between any pair of islands. When the population is considered as a whole a more consistent pattern of relatedness emerges. In this case St Mary's is most closely associated with Cornwall and Devon, and next with the outside world. A second cluster is formed by Bryher associating with Tresco after 10 generations and these merging with St Agnes after 11 generations. The two main clusters merge after 12 generations but St Martin's does not become homogeneous with the rest of the group until a minimum of 13 generations have elapsed. These are the sorts of associations that would be expected on the basis of the geography and the history of the Isles. St Mary's, being the centre of trade and commerce, and containing the only port for the islands, would be expected to be most closely associated genetically with the wider world outside. Tresco and Bryher, on the basis of their close geographical proximity, would be expected to share considerable common ancestry.

The stochastic model

The results of the analysis of kinship derived from migration patterns are shown in Fig. 4.4. This shows the development of kinship for the periods 1750–1799, 1800–1849, 1850–1899 and the combined period 1750–1899. Mean kinship within populations is plotted by generation. The progressive breakdown in isolation of the population over time is illustrated by the lower mean kinship and the more rapid approach to equilibrium values in successive periods. The high rates of endogamy and the small amount of long-range immigration, combined with drift caused by relatively small effective population sizes, have produced high mean kinship values, particularly for the earliest period. When the total population is considered, mean kinship is lowest, and equilibrium is approached very rapidly. The mean kinship after 50 generations for each time period 1750–1799, 1800–1849 and 1850–1899 is 0.0795, 0.0263 and 0.0109 respectively. These values are very high. However, when the population is considered as a whole, mean kinship after 50 generations is 0.0077, exactly the same as the equilibrium value quoted for Barra in the Outer Hebrides (Morton *et al.*, 1976), and within the range of kinship values derived from pedigree and isonymy studies. The high kinship values for the earlier time periods appear to be due to the low systematic pressure of migration from the outside world, combined with the high rates of island endogamy. The populations are also small, providing the conditions for large drift effects. Both the migration from the outside world and same-island endogamy decrease over time, with a concomitant increase in inter-island exchange within the Isles of Scilly. The effects of this pattern are clearly illustrated in the patterns of developing kinship under the model.

Table 4.5 gives the coefficients of hybridity for each pair of populations for each time period. The most obvious feature of these coefficients is their lack of consistency over time. For example, Tresco and St Agnes have the largest coefficient in 1750–1799, the smallest coefficient in 1850–1899 and an intermediate coefficient in 1800–1849. Bryher and St Martin's, which are highest in 1850–1899, have the third lowest value for 1750–1799. However, a trend towards decreasing hybridity with time is

84   *P. D. Raspe*

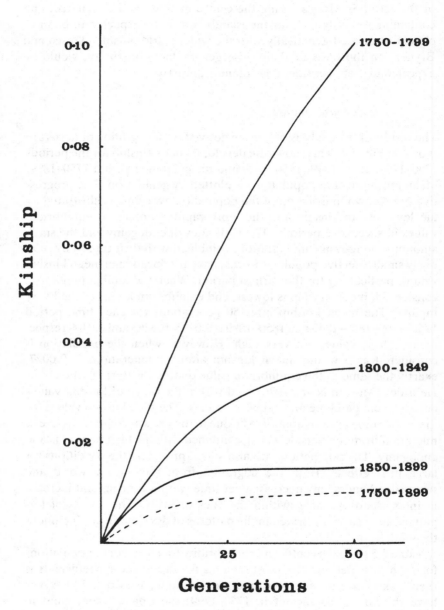

Fig. 4.3. Kinship from migration matrices.

Table 4.5. *The coefficients of hybridity for all pairs of islands for each time period*

	1750–1799	1800–1849	1850–1899	1750–1899
SM–TR	0.0089	0.0037	0.0027	0.0010
SM–SA	0.0217	0.0088	0.0030	0.0021
SM–MT	0.0149	0.0105	0.0030	0.0022
SM–BY	0.0221	0.0094	0.0037	0.0025
TR–SA	0.0310	0.0045	0.0018	0.0013
TR–MT	0.0210	0.0059	0.0030	0.0016
TR–BY	0.0246	0.0044	0.0029	0.0013
SA–MT	0.0070	0.0041	0.0036	0.0015
SA–BY	0.0071	0.0043	0.0032	0.0015
MT–BY	0.0073	0.0062	0.0051	0.0020

See Table 4.1 for abbreviations.

apparent. The highest coefficient of hybridity in 1850–1899 is smaller than the lowest coefficient in 1750–1799. The higher coefficients in 1800–1849 tend to overlap with the lower coefficients in the preceding period, and the lower coefficients are within the range of the highest coefficients for the subsequent period. However, for each pair of islands the coefficient of hybridity decreases consistently over time. The values for the combined populations tend to be the lowest of all. In each case the value for the combined population is lower than for the period 1850–1899, although the pattern is again different.

Other Studies

A number of other populations have been analysed using the deterministic model of Hiorns *et al.*, notably the Otmoor villages in Oxfordshire (Hiorns *et al.*, 1969), contemporary Reading in Berkshire (Coleman, 1980) and the Island of Karkar, New Guinea (Harrison, Hiorns & Boyce, 1974). The time taken to attain 95% homogeneity in the Isles of Scilly is most comparable to that reported for the 8 parishes of Otmoor where the figures are 25 generations in the period before 1850 and 12 generations for the period after 1850. This compares with the maximum and minimum figures for Scilly of 28 generations to 95% homogeneity in 1750–1799 and 10 generations in 1850–1899. However, the patterns by which such relationship develops are rather different in the two populations, at least during the early period. In the Otmoor analysis for the period before 1850, the first populations do not merge until 15 generations have elapsed, but total homogeneity is attained after a further 10 generations. In the case of the Isles of Scilly for the earliest period, the first populations merge after only 9 generations, and gradually continue to become more and more homogeneous over the next 19 generations. The differences in the later periods are not marked, partly because of the smaller generation depth. These differences seem to be mainly because in the Isles of Scilly common ancestry develops primarily as a result of inter-island marital exchange, while in the Otmoor case such relatedness is primarily due to the effect of migration from the outside world. This is demonstrated by the fact that when the outside world was disregarded in the Otmoor study, the length of time to attain 95% homogeneity was vastly increased, reaching 142 and 207 generations for the early and late periods respectively. In the Isles of Scilly, the exclusion of the small number of marriages with the outside world made very little difference to the length of time required to reach 95% homogeneity.

In the Reading survey, a slightly modified version of the model was used in order to accommodate data obtained by different methods. The records used in this survey encapsulate the movement patterns of a contemporary urban population. In this case few generations were required for the population to reach 95% homogeneity. Examination of Greater Reading, including marital contributions from the outside world, showed that a maximum of 7 generations was required for the population to share 95% relatedness, and that all but one of the populations were so related after only 5 generations. The first subpopulations became homogeneous after only 2 generations. This again appears to be primarily due to the high rate of immigration from the outside world, which averaged over 50%. When the population was considered as a closed system, including the contributions from 11 areas in the surrounding region but excluding the outside world, a much longer time was required to bring the population to 95% homogeneity. In this case a maximum of 15 generations was required, but the majority of the subpopulations shared common ancestry within 9 generations. When the outside world was included, the maximum time to homogeneity dropped to 7 generations. All these figures are lower than the comparable figures for the Isles of Scilly, reflecting the greater degree of isolation of the island populations.

The Karkar analysis demonstrated an unusual pattern of migration determined by geographic and social barriers, where, because of the largely inaccessible nature of the volcanic centre of the island, marital movement was around the periphery. The effect of village propinquity was marked, but the major factor determining movement patterns was the division of the population into two linguistic groups, between which there was very little exchange. The minimum length of time to reach 95% shared ancestry for all populations on the island was 90 generations under the open model, and 110 generations when the outside world was excluded. These high figures may be partly due to the clustering of nearby villages in order to produce migration matrices of manageable size, but are attributed to the high rates of endogamy with respect to village clusters.

The results produced under the stochastic model of Imaizumi *et al.* (1970) can also be compared with those derived for other populations using the same model. The mean kinship values for the Isles of Scilly after 50 generations are rather higher than most other populations which have been reported, and are most comparable to those reported for Karkar Island (Boyce *et al.*, 1978). In this study the population was divided into age cohorts in order to investigate a secular trend in migration patterns, and the mean kinship values after 50 generations of simulation were 0.0126 (15–29 years), 0.0151 (30–44 years) and 0.0245 (45 and older). The values for the nineteenth century in Scilly are similar: 0.0263 in 1800–1849, and 0.0109 in 1850–1899. However, mean kinship after 50 generations in the earliest period is 0.0795, higher than reported anywhere else. This may be partially due to inadequacies in the data for the very earliest period, where information on migrants might not have survived. However, the trend to decreasing kinship over time, which appears to result from the increase in movement between populations and the associated increase in migration from the outside world, is clearly marked.

When the Isles of Scilly are compared with other British populations, the high degree of isolation becomes more apparent. Relethford (1980) has investigated the effects of changing population size in western Ireland using the model and found mean kinship levels of 0.0065 after iteration for 50 generations. The build-up of kinship was rapid so that mean kinship of 0.0050 was reached after only 10 generations. Similar values were found for the island of Barra in the Outer Hebrides where equilibrium values for mean kinship of 0.0077 were reached after 122 generations, although these values were approached much sooner. The mean kinship values for the combined population of the Isles of Scilly, where effective population sizes are much larger than in the individual time periods, tend to fall within this range. Roberts, Jorde & Mitchell (1981) have reported significantly lower mean kinship for a sample of the population of Cumbria. However, the discordance of the mean kinship values of 0.000020 produced by the model and estimates based on genetic data of 0.0023 throw some doubt on the validity of the comparison with the present results.

In general, the analysis has confirmed the high degree of isolation of the
Isles of Scilly which has been reported elsewhere by isonymy methods
(Raspe & Lasker, 1980) and pedigree inbreeding (Raspe, 1982). In both
cases the Isles of Scilly tended to group with highly isolated populations
such as Tristan da Cunha, religious isolates and remote hamlets in the
Italian Alps, rather than with other British populations.

References

Ashbee, P. (1974). *Ancient Scilly: From the First Farmers to the Early Christians.*
Newton Abbot: David & Charles.

Beckman, L. (1980). Time trends in endogamy rates in northern Sweden. In
Population Structure and Genetic Disorders, ed. A. W. Eriksson, H. R.
Forsius, H. R. N. Nevanlinna, P. L. Workmann & R. K. Norio, pp. 73–80.
London: Academic Press.

Bodmer, W. F. & Cavalli-Sforza, L. L. (1968). A migration matrix model for the
study of random genetic drift. *Genetics*, 59, 565–92.

Borlase, W. (1756). *Observations on the Ancient and Present State of the Isles of
Scilly.* Oxford: W. Jackson.

Bowley, R. L. (1968). *The Fortunate Islands: A History of the Isles of Scilly.*
Reading: Bowley Publications.

Boyce, A. J., Harrison, G. A., Platt, C. M., Hornabrook, R. W., Serjeantson,
S., Kirk, R. L. & Booth, P. B. (1978). Migration and genetic diversity in an
island population: Karkar, Papua New Guinea. *Proceedings of the Royal
Society London, B*, 202, 269–95.

Boyce, A. J., Holdsworth, V. M. L. & Brothwell, D. R. (1972). Demographic
and genetic studies in the Orkney Islands. In *Genetic Variation in Britain*, ed.
D. F. Roberts & E. Sutherland, pp. 109–27. London: Taylor and Francis.

Clegg, E. J. (1975). Marriages in Lewis and Harris 1861–1966. In *Biosocial
Inter-relations in Population Adaptation*, ed. E. S. Watts, F. E. Johnson & G.
W. Lasker, pp. 147–63. The Hague: Mouton.

Coleman, D. A. (1980). Some genetical inferences from the marriage system of
Reading, Berkshire and its surrounding area. *Annals of Human Biology*, 7,
55–76.

Gill, C. (1975). *The Isles of Scilly.* Newton Abbot: David & Charles.

Harrison, G. A., Hiorns, R. W. & Boyce, A. J. (1974). Movement, relatedness
and the genetic structure of the population of Karkar Island. *Philosophical
Transactions of the Royal Society of London, B*, 268, 241–9.

Heath, R. (1750). *A Natural and Historical Account of the Isles of Scilly.* London:
Manby & Cox.

Hiorns, R. W., Harrison, G. A., Boyce, A. J. & Küchemann, C. F. (1969). A
mathematical analysis of the effects of movement on the relatedness between
populations. *Annals of Human Genetics*, 32, 237–50.

Imaizumi, Y., Morton, N. E. & Harris, D. E. (1970). Isolation by distance in
artificial populations. *Genetics*, 66, 569–82.

Malécot, G. (1948). *Les Mathématiques de l'Hérédité.* Paris: Mouton.

Models of human migration

Malécot, G. (1950). Quelques schéma probabilistes sur la variabilité des populations naturelles. *Annals Université Lyon, Science A*, 13, 37–60.
Morton, N. E., Smith, C., Hill, R., Frackiewicz, A., Lew, P. & Yee, S. (1976). Population structure of Barra. *Annals of Human Genetics*, 39, 339–52.
Morton, N. E., Yee, S., Harris, D. L. & Lew, R. (1971). Bioassay of kinship. *Theoretical Population Biology*, 2, 507–24.
Raspe, P. D. (1982). *The Mating Structure of a Subdivided Population* Ph.D. Dissertation (unpublished).
Raspe, P. D. & Lasker, G. W. (1980). The structure of the human population of the Isles of Scilly: inferences from surnames and birthplaces listed in census and marriage records. *Annals of Human Biology*, 7, 401–10.
Relethford, J. H. (1980). Simulation of the effects of changing population size on the genetic structure of Western Ireland. *Social Biology*, 27, 53–61.
Roberts, D. F., Jorde, L. B. & Mitchell, R. J. (1981). Genetic structure in Cumbria. *Journal of Biosocial Science*, 13, 317–36.

5   *Rural-to-urban migration*

BARRY BOGIN

The movement of people from place to place is a major determinant of the biological structure of human populations. Such movement injects new genetic, physiological, and morphological variability into the recipient populations. It may also deplete these sources of biological variation from the non-migrating donor population. Rural-to-urban migration is one of the most prevalent types of human movement. The extent of this type of migration in today's world is enormous. In 1800, there were about 25 million people living in urban areas. In 1980, there were about 1.8 billion. By the year 2000, it is estimated that this number will rise to 3.2 billion, a 128-fold increase in two centuries (Rogers & Williamson, 1982). In contrast, the natural increase in total world population will only be 6.4-fold in the same two centuries (one billion people in 1800 to 6.4 billion in the year 2000).

This chapter reviews the biological consequences of rural-to-urban migration for human populations. The study of the process of rural-to-urban migration and its biological effects is important for four reasons. First, it entails movement into a habitat and an ecological niche – the city – that is evolutionarily novel for our species. Secondly, it is the most common type of migration that has occurred in all periods of recorded history (Smith, 1984). Thirdly, it is occurring more rapidly today than ever before, especially in the least developed nations of the world (United Nations, 1980). Fourthly, we understand relatively little about the long-term effects of this migration on human biology.

Following from these four reasons for the study of rural-to-urban migration, this review begins with a description of the extent of migration today and in the past. This discussion leads into a consideration of the possible consequences of living in the urban environment on human biology. The following section divides the information on rural-to-urban migration into four areas. These are: (1) growth and development, (2) fertility and demography, (3) morbidity and mortality, and (4) the biology of the rural non-migrating population. The issue of biological selection of urban migrants, e.g. 'do only taller, less fertile people migrate?' is discussed, as appropriate, in each of these areas.

90

The literature covered is for voluntary migrations only. Forced migrations due to natural disasters, humanly caused disasters, social and political repression, and war require separate treatment. The genetic consequences of urban migration are not discussed, since adequate coverage of this topic appears elsewhere (see Boyce, 1984).

Finally, a tentative explanation for the patterns of biological change and adaptation to the urban environment is offered. This tentative explanation may serve as the basis for developing testable hypotheses that will allow us to predict and, where necessary, alter the long-term biological consequences of rural-to-urban migration.

The extent of urbanization

Let us first consider the extent of rural-to-urban migration throughout the historic period and even back to its archaeological context. It is logical to assume that urban migration began at the time when cities came into existence. This is not true, however. Sometime before 3000 BC, urban centres appear in ancient Sumer. These 'cities' developed from villages, which in turn had grown from small agricultural settlements. Though these cities represented 'a new magnitude in human settlement' (Childe, 1942, p. 94), their growth was slow and they were still considered as an extension of the agricultural settlement. The Sumerian language makes no distinction between village, city, or 'any permanent cluster of houses made of sun-dried mud bricks' (Tuan, 1978, p. 2). Farming areas around the city were called *uru.bar.ra*, the 'outer-city' (Oppenheim, 1974). Though people may have moved from the outer-city to the inner-city, such movement cannot properly be called rural-to-urban migration by today's definition of this term.

Ancient and mediaeval cities of Europe and Asia continued the practice of mixing settlements of houses, government buildings and religious edifices with orchards, vineyards and gardens. Urbanites in these cities still practised a rural lifestyle on a daily basis (Tuan, 1978). A clear distinction between city and countryside and a clear conception of rural-to-urban migration is not apparent in historical records or in literature until the late seventeenth century.

Population size of the mediaeval city was small by modern standards. The typical size of the city ranged from 2000 to 20 000 inhabitants until the end of the sixteenth century (Mumford, 1956). Limited urban population growth may have been determined by two factors, disease and rural in-migration. McNeill (1976, 1979) takes a novel approach to history and migration studies. He views much of history and most of migration as driven by disease. A concentration of large and dense human populations in an urban area establishes the conditions for the communication of infectious disease from one person to another. Settled village life had set up the conditions for the infection of anthrax, brucellosis, and tuberculosis in human populations. Urban life increased the chances for the spread of more devastating epidemic diseases such as cholera, smallpox and plague (Cockburn, 1967; Armelagos & Dewey, 1970).

McNeill argues that epidemic die-off in the urban centres of the world prior to about 1750 did not allow cities to sustain their own populations. Storey (1985) estimated the birth rate and the age at death for a skeletal sample from the pre-Columbian urban centre at Teotihuacan. The sample was excavated from the trash midden burials of an 'apartment compound' occupied by low socioeconomic status residents of the archaeological city. The estimate of mortality prior to age 15 is relatively high at 50%, and one-third of these deaths occurred during the first year of life. Life expectancy was low: most adults only lived to their late thirties (87% were estimated to have died by age 40). Storey shows that this rate of mortality is similar to that of pre-industrial Old World cities such as Rome and London. The estimates of mortality for the low socioeconomic class of Teotihuacan belie the claims for greater longevity for pre-Columbian urban populations based on skeletal remains of the ruling class. Storey adds that excavations at the apartment complex site indicate that the poor sanitation, high population density and problematic food supply of the Old World cities seem to have been characteristic of New World cities too. Thus, believes Storey, the McNeill model of in-migration to maintain urban populations is valid for Teotihuacan. Based upon the skeletal sample she and others studied, population size rose in the early period of the city but remained constant during the middle and later periods because, although more deaths than births occurred, in-migration made up the difference.

Kennedy (1973) presents data showing that even as late as 1850 English and American cities had mortality rates higher than their rural areas. Rural in-migration was necessary to maintain urban population growth. Why people would want to move into an urban death-trap is unclear. McNeill states that urban growth depended on 'moral and property systems, family practices, and biological/technological balances in the countryside that allowed and encouraged raising more children than were needed . . .' (1979, p. 96). Thus, the rural 'push' of overpopulation and socioeconomic limitations were major factors. This was the case for much of the emigration from the Irish countryside. Large families, exclusion of all but the first-born son from inheritance of land, and the lack of rural opportunities for wage-earning employment resulted in massive urban migration prior to the famines of 1845–1848 (Kennedy, 1973). Even today, these reasons, plus the urban 'pull' of an expected higher standard of living in the city, are sufficient to explain rural-to-urban migration in the least developed countries. These urban migrants believe life will be better, even as they move into an urban shantytown with high rates of unemployment and poverty (Rogers & Williamson, 1982). Given this history, it is probable that the phenomenon of rural-to-urban migration began when people realized that rural and urban life presented the individual with different opportunities. Some of these opportunities directly related to human biology in the areas of fertility, disease and mortality.

Cities, as we know them today, came into existence at the end of the sixteenth century. By the time of Shakespeare, London was a city of some 100 000 people living within an area of one square mile. Massive urban in-migration from rural areas was one consequence of this degree of population concentration. Because of epidemic deaths within the city, London required on average an annual rate of in-migration totalling about 5000 people in order to sustain its population (McNeill, 1979). Though this rate of migration pales before the current rates of movement to some cities (Mexico City receives about 5000 immigrants per day), it is still an impressive number.

By about 1750 the rate of urban mortality due to epidemic disease was sharply reduced. This may have been due to the dual effects of biological adaptation of the human host and pathogen towards each other, as well as improved sanitation, water treatment, and other public health measures (Dubos, 1965). Urban populations no longer required in-migration from the countryside to be sustained, but rural-to-urban migration continued unabated. The political, economic, and social practices of mill owners in the cities and landowners in the countryside in the pre-1750 period were still 'pushing' people from the country and 'pulling' them into the city in the two centuries that followed.

Between 1750 and 1950, the bulk of rural-to-urban migration took place in what are now the developed nations. There was also a great deal of international migration, especially from the Old World to the Americas. Many of these migrants moved from the rural villages of the 'Old Country' to the cities of the New World. By 1950, 53% of the population of the more developed nations were urbanites, compared with only 16.7% of the population of the less developed countries. By the year 2000 the percentage for all less developed countries will still not exceed 50% (44.5% is the estimate), but the developed nation figure will be

about 79% (United Nations, 1980). Therefore, the biological consequences of living in cities, as we know them today, may be more readily
observable in the populations of the developed countries. We can also use
the past experience of the developed nations to construct explanations
and hypotheses about present-day urban migration that may be tested
with the more recent data from the less developed countries.

Most of the post-World War II increase in rural-to-urban migration has
occurred in the less developed countries of the Third World. Since 1975,
for the first time in history, the majority of the world's urban population
lives in the cities of the Third World nations (Rogers, 1982). Of the 1.8
billion urban dwellers of 1980, 1.1 billion of them live in these cities. This
number represents 34% of the total population of the less developed
nations. In 1960 only about 26% of the population in the less developed
countries lived in urban areas (World Bank, 1981). This 31% rise in urban
population translates into 260 million more people living in the cities than
if the 1960 percentage had remained constant. By the end of this century,
migration and natural increase will result in three-quarters of the world's
20 largest cities being located in the less developed countries (United
Nations, 1980).

Large populations, dense populations, and high rates of in-migration
are the most important characteristics of cities today. These characteristics lead to several types of problems for humans. These include the '. . .
technological problems of water supply, pollution, waste management;
socio-economic problems of poverty, unemployment, and social conflict;
and biological problems of disease and ill health' (Harrison & Jeffries,
1977, p. 65). It is to these biological problems that we now turn our
attention.

Are cities good or bad for people?

There are several different lines of reasoning that lead us to suspect, *a
priori*, that urban living should be deleterious to human biology. This
evidence includes a consideration of:

(1) *Evolution.* Humans evolved as nomadic hunters and gatherers, living
in small band populations. Cities are composed of sedentary, industrially
and technologically employed peoples living in large groups. Humans are
capable of a great range of adaptive responses to new environmental
stress, but the genetic limits of this range are determined by the nature of
adaptation to past environments (Harrison & Jeffries, 1977). The new-
ness of the urban environment in evolutionary time is, therefore, a
potential threat to human physical well-being.

(2) *Development.* People develop phenotypic adaptations to their local environments. These include the irreversible changes in growth that occur during childhood and adolescence as well as the development of disease immunities that occur even in adulthood (Weissman, Hood & Wood, 1978). Adaptation to local diets and activity patterns also shape human physiological adaptations. Migration to an urban environment following long-term residence in a rural area may well present a significant stress to human physiology.

(3) *Biosocial adaptation.* Band, tribal and chiefdom societies were the basis of human social organization for the 99% of our evolutionary history that preceded the appearance of the first cities. These social groups are characterized by kinship (biological and fictive) as the organizing principle of society. However, kinship is less of a determinant of urban population structure. The large, densely populated conglomerates of cities require different patterns of social organization. These patterns have been analysed in the social science literature dealing with urbanization (Hauser & Schnore, 1965; Michelson, 1970; Butterworth & Chance, 1981). Suffice it to state here that rural-living peoples continue to organize their lives along the lines of our pre-urban ancestors. When these rural peoples migrate to the city they may be forced into rapid social change. This may lead to considerable psychological stress, and precipitate physical and mental illness (Carlstram & Levi, 1971; Clegg & Garlick, 1981).

Thus, rural-to-urban migration may lead to a considerable reduction in the 'fitness' of human populations, as reflected in reduced fertility, poor growth and development, poor health, and greater mortality at all ages.

The prevalence of disease and mortality in the pre-modern city has already been discussed. but what of the modern city? In an essay on the historical development of the modern city, Tuan (1978) explores 'the idea that cities are artifacts and worlds of artifice placed at varying distances from human conditions close to nature'. It is Tuan's opinion that the natural human condition is that 'bound to food production and . . . the natural rhythms of day and night and of the seasons'. To Tuan the development of contemporary cities, such as New York, Los Angeles and London, is not inevitable, natural or desirable.

Life in the cities of the poorer nations is viewed with even greater disdain by O'Dell (1984). In her opinion, the urban poor of the less developed countries live in 'crowded, squalid, often temporary dwellings [that] sprawl along mud paths littered with garbage, human refuse, and pests of every variety . . . Urban mortality rates, especially for small children, are higher than those in rural areas'.

The opinion of other writers is that the evils of the city even extend back into the rural areas. Lipton (1977) perceives an urban bias in public policy. Keyfitz (1982) describes this bias as 'the perversity of urbanization'. By this he means that urban centres prosper at the expense of rural social and economic development. As the standard of living in the countryside deteriorates this 'pushes' people into the city. The result is overurbanization, a force of unknown magnitude and significance for human survival.

To counteract the pernicious effects of overurbanization, some researchers write about 'optimal city size' and place that size at 250 000 to 500 000 people (Rogers & Williamson, 1982). There is little hard evidence that large modern cities are, in themselves, harmful to people. These 'optimal' numbers are calculated in terms of economic factors, not in terms of social or biological realities. The reality is that 'life in the city implies first and foremost a regular supply of goods and services, and the existence of institutions associated with this provision' (Tanner & Eveleth, 1976, p. 145). In the countryside of the less developed countries, the rhythm of nature means that prior to the harvest people may go without adequate food. Water may be contaminated by disease organisms, especially in the rainy season. Work loads are heavy, even for children and pregnant or lactating mothers. Social, economic and health support services are rare. In the cities of these developing nations, food storage and distribution services provide a supply of food throughout the year. Treated water is available. The physical demands of labour are often less than those of rural agriculture. The largest health, educational and welfare facilities are in the city. In short, the city may not be such a bad place.

Of course, not all of the people of the city share equally in these services and benefits. But the slums and suffering that exist in the city are not the fault of urbanization *per se*, they are the fault of our whole society. Perhaps at this point it is best to counter the view that cities are bad for people with the notion that the city 'is the product of human society and arises from human social nature' (Briggs, 1983, p. 371). The city is not artificial. As we will see in the sections to follow, the biological consequences of living in the city are not deleterious. Most empirical studies of rural-to-urban migration find that the health of migrants is better, their mortality is lower, their children grow taller, and their fertility is lower than for rural sedentes. By evolutionary, developmental and biosocial criteria, both rural-born and city-born people usually adapt successfully to the urban environment.

Review of the literature

The human sciences were slow to appreciate the importance of this new phenomenon of rural-to-urban migration. It was not until 1885 that Ravenstein published his famous 'laws of migration'. Interest in the biological effects of urban migration followed with Livi's (1896) study of the growth of Italian migrants and Ammon's (1899) work on the growth of rural-born Germans in Baden. Migration studies were of sporadic interest through the 1960s. In the late 1960s the International Biological Programme was established, and one of its aims was to consider human biology in various environments. Soon after, edited works by Boyden (1970), Harrison & Gibson (1976) and Baker (1977*a*) appeared that addressed the issues related to the human urban environment. The remainder of this chapter reviews the literature on the biological consequences of rural-to-urban migration.

Growth and development

The earliest studies of the biology of urban migrants deal with physical growth. So, it is appropriate to begin the review with these works.

Measures of physical growth and development are often used to assess the overall quality of the environment in which children live and the health and nutritional status of the population of which they are members (Tanner & Eveleth, 1976; Waterlow *et al.*, 1977). Meredith (1979) and Malina *et al.* (1981) review studies from the period 1880–1920 that show rural-living children in the United States and Europe were taller than their urban peers. For example, Steegmann (1985) found in a study of eighteenth century British military records that the stature of rural-born recruits averaged 168.6 cm and the urban-born recruits averaged 167.5 cm, a statistically significant difference. By 1930 this pattern was reversed and urban children were consistently taller and heavier than rural children. These studies give us some ideas about changes in the rural and urban environment at the turn of the century. Rural versus urban differences in growth (height, weight, body proportions), body composition (lean and fat body mass), and maturation (skeletal age, menarche) have been summarized by Tanner & Eveleth (1976), Meredith (1979) and Susanne (1984). Despite dozens of studies on urban versus rural populations there are only a few that treat rural-to-urban migrants.

Boas (1912, 1922) cites studies by Livi (1896) and Ammon (1899) that treat urban migrants. Livi found that the children of urban migrants in Italy were taller than rural sedentes. He believed the reason for this was heterosis, the marriage of urban migrants from different rural regions leading to 'genetic vitality' in their offspring. Ammon also found the children of migrants to be taller than rural sedentes, but he argued for the action of natural selection to explain this. Perhaps he meant that in the rigours of the urban environment only the 'fittest' (the tallest?) would survive.

Livi's and Ammon's speculations that heterosis or natural selection were at work stem directly from their erroneous belief that human types were genetically fixed and that types would not change when exposed to different environments. That belief was shattered by the publication of Boas' (1912) study of the 'Changes in the bodily form of descendants of immigrants'. Boas found that the children of immigrants to the United States were taller and shaped differently from their parents and from the non-migrating populations from which their parents came. He stated that neither natural selection nor heterosis could adequately account for these changes. Rather, modifications in the process of growth and development as a response to environmental change were responsible. Boas was vigorously attacked for this position. In a series of papers he whittled away at his attackers (see Boas, 1940). His evidence against the fixity of types was that: (1) the physical differences between parents and children appear early in the life of the child and persist until adulthood, (2) the longer the childhood exposure to the American urban environment the greater the physical difference, (3) children from large families are shorter than children from smaller families of the same 'racial type', and (4) differences between parents and children are greater when both were foreign born (meaning that only the children would have been exposed to the new environment during the developmental years) than when the child was American born (meaning that the parents may have spent some of their growth years in the United States).

The now classic studies of Shapiro (1939) on the growth of Japanese children in Japan and Hawaii, and of Goldstein (1943) and Lasker (1946, 1952) on the effects of growth of Mexicans and Chinese in the United States, confirmed the nature of human developmental plasticity. Today, human developmental plasticity is taken for granted, but it was the migration data of Boas that established the validity of this phenomenon.

Shapiro's Japanese migrant study compared the growth of Hawaiian-born Japanese of Japanese immigrant parentage, Japan-born Japanese who migrated to Hawaii, and Japanese sedentes living in the same villages from which the migrants originated. The sedentes were mostly farmers and labourers in rural villages. The exact location of the migrants is not given, but 65% of the recent immigrants were employed in sedentary occupations (store clerks), and 76% of the Hawaiian-born were either students or sedentary workers. The immigrants appeared to have been developing an urban lifestyle. The sedentes and the recent immigrants differed in a few anthropometric measurements, some increasing and some decreasing with migration. The largest differences were between the immigrants and the Hawaiian-born. The latter are taller and more linear in body build than their parents or the sedentes. Shapiro argues that after migration there were improvements in diet, health care, and socioeconomic status and that these conditions, associated with an urban lifestyle, are responsible for the growth changes.

A nearly ideal study of rural-to-urban migration and growth was carried out in Poland. The results were published in Polish by Panek & Piasecki (1971) and were summarized in English by Tanner & Eveleth (1976). A new industrial town was created on the outskirts of Cracow in 1949. In 1965 the population reached 1599 persons per square kilometre, similar to many European cities. Most of the population growth was due to migration from rural villages. The children and youths measured for the study had been born in the new city or had lived there at least 10 years. The children living in the 'new city' were, on average, five to six centimetres taller and one to two kilograms heavier than the rural sedentes. The urban children matured earlier, both for tooth eruption and age of menarche. Thus, even in the post-World War II period, the children of urban migrants experienced significant improvements in growth.

Children living in the cities of the less developed countries are usually taller and mature earlier than their rural age peers. But, migrants to urban slums in these countries do not usually experience the benefits of the urban environment. In Asia, Africa and Latin America these slums are often on the outskirts of the cities. As squatter settlements they have no official access to city services and facilities. Not surprisingly, the growth of migrant children living in these slums is not significantly different from that of children living in the impoverished rural areas (Morley *et al.*, 1968; Johnson, 1970; Villarijos *et al.*, 1971; Davies, Mbelwa & Dore, 1974); Graham *et al.*, 1979). In one case, the children living in an urban slum, children of Oaxaca, Mexico are significantly shorter and lighter than nearby rural children from a reasonably prosperous town (Malina *et al.*, 1981).

In contrast to these cases, Bogin & MacVean (1981) studied the growth of three groups of children living in Guatemala City: (1) children of two city-born parents, (2) children of two rural-born migrant parents, and (3) children of one city-born and one rural-born migrant parent. All of the children were of low socioeconomic status and attended a public school. Most lived in impoverished areas of the city, but none were from the types of urban slums discussed above. All of these children were taller and heavier than rural-living children of the same age. Also, as expected, children of city-born parents were significantly taller than children of rural-born parents. Unexpectedly, children with one city-born and one rural-born parent were the tallest. This was true for families in which the mother or the father was the migrant. The families of the rural-born parents were of significantly lower socioeconomic status than the families of the two other kinds of parents. This difference probably correlated with factors responsible for the growth differences, but other important information, such as age of the mothers and age at migration to the city, were not known.

Johnston *et al.*, (1985) studied a sample of people from a resettled community in Guatemala City. The community, called El Progreso in this study, was established by the Guatemalan government after the earth-quake of 1976, and became the home for 10 000 disadvantaged people from rural villages and neighbourhoods in Guatemala City that were destroyed. In Table 5.1 are presented the heights and weights of several samples of children, aged seven years old, from Guatemala. It is difficult to interpret these data in relation to migration since the migration history of the children and adults of El Progreso were not reported. However, these results show that child growth is a sensitive indicator of environmental quality. The children of El Progreso were shorter and lighter than all other urban samples of children, attesting to the disadvan-taged socioeconomic status of their new community. They were slightly taller than children living in rural villages, but lighter than rural Indian children. To the extent that height represents the long-term environmen-tal influences on growth and weight represents the shorter-term influences (Waterlow *et al.*, 1977), these comparisons may mean that the El Progreso children had suffered some recent growth delays (caused by the earth-quake and resettlement perhaps), but experienced a life-long environ-ment for growth that was better than that found for children who still lived in rural villages.

A possible reason for taller children in this Guatemalan study is selective migration. Such selection may explain the biological correlates of migration in some of the other studies already discussed. By selective migration is meant that migrants may not be a random sample of the rural population to which they had belonged. Migrants may be genetically taller, mature more rapidly, or differ in other physical aspects from rural sedentes. Positive assortative mating between such people may further differentiate migrants from sedentes.

Table 5.1. *Heights and weights of five samples of seven-year-old Guatemalan children*

Sample	Height		Weight	
	Mean	SD	Mean	SD
Boys				
High SES[1]	120.3	4.9	24.0	4.6
Low SES[2]	113.4	5.3	20.5	2.5
El Progreso[3]	111.6	5.5	19.2	3.1
Rural Indian[4]	110.9	4.5	19.6	2.8
Rural Non-Indian[5]	110.4	4.4	18.7	1.9
Girls				
High SES	120.0	5.3	24.1	4.2
Low SES	112.2	5.4	19.5	2.3
El Progreso	111.4	5.7	19.0	2.8
Rural Indian	110.2	4.8	20.0	2.2
Rural Non-Indian	109.6	3.9	18.1	1.7

[1] Johnston *et al.* (1973)
[2] Bogin & MacVean (1978)
[3] Johnston *et al.* (1985)
[4] Bogin & MacVean (1984)
[5] Yarbrough *et al.* (1975)
Abbreviations: SD, standard deviation; SES, socioeconomic status.

Several studies provide tentative support for biological selection. In Shapiro's (1939) study, the recent Japanese immigrants to Hawaii were taller than the Japan sedentes. However, age at migration, length of time in Hawaii, and pre-migration living conditions were not known. Though Shapiro suggested that selective migration was possible, he argued more strongly for the plasticity of human growth in the new environment. By this he meant that migrant children grew more than sedentes, but only after they migrated to Hawaii. Steegmann's (1985) historical study of eighteenth century British military recruits found that conscripts who migrated from their county of birth were significantly taller (169.1 cm) than recruits living in the county of their birth (167.6 cm). Recall though that he also found that urban-born recruits were shorter than rural-born men. Steegmann points out that eighteenth-century Britain was a developing country and that urban areas were characterized by food shortages and unhygienic living conditions. Thus migration to the city was probably not responsible for an increase in stature; rather, taller men living in rural counties were more likely to migrate than shorter men.

Whether these were genetically taller men, or individuals who had experienced better living conditions prior to migration, is not known.

In the United Kingdom, Martin (1949) found that among men inducted into the army: (1) migrants (men living in a county other than their county of birth) were taller and heavier than the national average, and (2) migrants were taller and heavier than the natives of the recipient counties. However, Martin also found that the sedentes of the recipient areas were taller and heavier than the sedentes of the donor counties and that migrants to 'tall' counties were taller than migrants to 'short' counties. These two facts suggested that the receiving counties in general, and the 'tall' counties in particular, had better living conditions than the donor or 'short' areas, but data for this and age at migration were not known.

Based on a retrospective study of 14 different ethnic groups, Kaplan (1954) found that migration was selective for physical type. Growth differences between migrants and sedentes were found too soon after migration to be due to an environmental change. However, no account of age of migration or the pre-migration environment was given. Illsley, Finlayson & Thompson (1963) studied migrants into and out of Aberdeen, Scotland. They found that all migrants were taller than rural or urban sedentes. Migrants had generally better health than sedentes. Migrant women had lower rates of low birth weight and perinatal death for their children. Finally, migrants were generally of higher socioeconomic status than sedentes.

The Illsley et al. study suggests what the true meaning of migrant selection may be: it is more likely to be selection for socioeconomic status than biological selection per se. More recent studies confirm this. Kobyliansky & Arensburg (1974) studied migrants from Russia and Poland to Israel. The migrants were of higher socioeconomic status and were taller than the Eastern European sedentes. Mascie-Taylor (1984) reviewed geographic and social mobility in England and found that the effects of selection are additive; migrants tended to be the taller individuals of any geographic area and the taller individuals within any social class. However, the higher social classes were also more mobile. Which is more important, stature or social class? Higher socioeconomic status can, by itself, lead to increased body size and rate of maturation. As Macbeth (1984) concluded in her review of this issue, the socioeconomic status difference confounds any unique biological difference between migrants and sedentes. Further confounding the issue is the fact that there is greater mobility to higher social classes for the tall (Tanner, 1969). However, the predominant selection of migrants seems to be for socioeconomic status rather than for tallness. This is supported by migration research from the social and economic literature (Rogers & Williamson, 1982) as well as the biological literature cited above.

Indeed, the most well controlled, prospective studies found no evidence for selective migration. These studies identified and measured individuals before migration. Lasker (1952, 1954) and Lasker & Evans (1961) found that age of migration and length of time in the new environment were responsible for growth differences between migrants from Mexico to United States and Mexican sedentes. Similarly, no growth differences were found between eventual urban migrants and rural sedentes in South Africa (De Villiers, 1971), Switzerland (Hulse, 1969) and Oaxaca, Mexico (Malina *et al.*, 1982).

In conclusion, it seems that the original work on the human growth response to migration of Boas (1912), Shapiro (1939), Goldstein (1943), Lasker (1952), and the more recent work summarized by Tanner & Eveleth (1976), Meredith (1979), and Malina *et al.* (1981) are correct. Environment, as mediated by such factors as socioeconomic status, is the primary determinant of biological change in growth and development following rural-to-urban migration.

Fertility and demography

There are two major questions relating to the demographic consequences of rural-to-urban migration. The first is: what is the most important reason for the phenomenal growth of urban populations? Rural-to-urban migration is certainly important, but the rate of migration sets only the level of urbanization. It cannot account for all of urban population growth. Reclassification of rural lands and peoples surrounding metropolitan areas also increases the urban population, but this has only a minimal effect (Rogers, 1982). Natural increase, i.e. the fertility of the urban population, seems to be the key factor. Natural increase sets the rate and the limit of urban population growth (Rogers, 1982).

The second question is: who are the more fertile, urban natives or the urban migrants? Most existing research finds that migrant women have lower fertility than rural sedentes (Myers & Morris, 1966; Zarate & Zarate, 1975; Orlansky & Dubrovsky, 1978; Rogers, 1982). Some studies find that migrants have lower fertility than urban natives (Orlansky & Dubrovsky, 1978; Rogers, 1982). Other research finds migrants have higher fertility than the urban-born (Hutchinson, 1961; Robinson, 1963; Zarate & Zarate, 1975; Baker, 1977*b*). Contradictory findings are not surprising since 'fertility' is the outcome of many biological, socioeconomic and cultural forces.

The data are consistent with regard to rural versus urban differences. Urban fertility rates (and mortality rates) are lower than rural rates everywhere in the world today. There are several factors associated with cities that have been cited for lowering women's fertility. These are:

(1) higher socioeconomic status (Orlansky & Dubrovsky, 1978),

(2) increased education (Kumudini, 1965; Macisco, Bouvier & Renzi, 1969; Chaundhury, 1978; Hinday, 1978; Ketkar, 1979; Lee & Farber, 1984),

(3) low infant and childhood mortality (Robinson, 1963; Frisancho, Klayman & Matos, 1976; Scrimshaw, 1978; Ketkar, 1979; Stinson, 1982),

(4) age at migration and marriage (Robinson, 1963; Macisco et al., 1969; Hinday, 1978),

(5) greater labour force participation (Freedman, Whelpton & Campbell, 1959; United Nations, 1975; Hinday, 1978; Ketkar, 1979; Lee & Farber, 1984).

(6) modernization and political liberalization (Goldscheider, 1971; Scholl, O'Dell & Johnston, 1976; Sabagh & Yim, 1980),

(7) a high personal achievement orientation and value system (Macisco, Bouvier & Weller, 1970; Benedict, 1972; Bogin & MacVean, 1981).

All of these factors are associated with urban living. However, some studies find that even after these factors (such as age, education, occupation) and other influences on fertility (such as parity) are controlled, urban living itself still accounts for a significant reduction in fertility (Myers & Morris, 1966; Rindfuss, 1976; Hinday, 1978; Lee & Farber, 1984).

Even though fertility in urban areas is lower than that in rural areas, there are a number of theoretical reasons why the fertility of rural-to-urban migrants might increase. The most important reasons relate to the opportunities for better health care, nutrition and social services in the city. Healthier people are likely to be either more fertile, or be able to carry a pregnancy to term and support an infant. Also, as shown in the previous section, the urban environment promotes faster growth and earlier maturation. All recent studies of rural versus urban physical maturation show that city-living girls reach menarche earlier than rural-living girls (Tanner & Eveleth, 1976). Presumably, the city folk also achieve functional fertility earlier. Finally, in many instances, more women than men migrate from rural areas to the city and most of these are young women. This is the case for the recent migrations in the less developed countries (Orlansky & Dubrovsky, 1978), and was true for migrations in Europe during the last century (Ravenstein, 1885). Thus, the potential for high migrant fertility is always present, but this potential is strongly countered by each of the factors listed above. Especially important are an average later age at first marriage, socioeconomic and cultural values that tend to delay births, and increased awareness and availability of contraceptive techniques, which all tend to reduce fertility among urban migrant women (Boyden, 1972).

These are the theoretical possibilities for urban migrant fertility; what are the actual data? Martine (1975) reviewed research from Latin America. He found that migrants had higher fertility than the urban-born. Families with one migrant and one city-born parent had intermediate fertility and within these families those with migrant fathers had higher fertility than those with migrant mothers. This suggests that the fathers' attitudes and desires for family size were at least as important as the mothers'. Martine also found that the migrants' contribution to urban growth is very large. He estimated that for the period around 1960, the total migrant contribution (including direct migration and natural increase among migrants) was 77% in San Jose, Costa Rica, 80% in Buenos Aires, Argentina and 91% in Bogota, Colombia.

Zarate & Zarate (1975) reviewed the research findings for migrant versus non-migrant fertility differentials. Their review included studies from the United States, Puerto Rico, Latin America, Asia, and Africa (only one African and three Asian studies were published at that time). They found that the fertility of migrant women was generally higher than that of urban natives, but lower than that of rural sedentes. An exception to the general trend was that younger migrant women, those under 35 years old, often had fertility rates lower than or equal to urban natives. Zarate & Zarate referred to studies from Puerto Rico by Macisco *et al.* (1969; 1970) and research in Thailand by Goldstein (1973) that may account for the age cross-over in fertility. These researchers speculated that the 'innovative character' and 'high achievement motivation' of the younger or more recent urban migrant may account for the decision to delay births.

Several additional studies have been published since these general reviews appeared. Liberty, Hughey & Scaglion (1976) and Liberty, Scaglion & Hughey (1976) compared the fertility of Omaha and Seminole Amerindian women living in rural and urban areas. Urban Omaha women both desired and gave birth to more children than their rural counterparts. This may have been due to the strong adherence to cultural traditions among these urban women. During the nineteenth and early twentieth century the Omaha underwent drastic population declines due to epidemic disease and poverty. High birth rate was one of the ways that the Omaha survived as a people. Attitudes favouring high birth rates may be ingrained in the culture. Also, since both rural and urban women lived at or below the poverty line, large families may have been viewed as a form of social security. In contrast, Seminole women in urban areas desired and gave birth to fewer children than their rural peers. Urban Seminole women were generally traditional in cultural practices, but they had more education and income than the rural women. The authors contended that the urban Seminole women may have been more susceptible to modernization and changing attitudes towards family size.

Rindfuss (1976) studied migrants from Puerto Rico now living in cities on the United States mainland. When women's age, education, initial parity and their husband's occupation were statistically controlled, there was no difference in current fertility between urban island residents, recent United States migrants, and long-term United States migrants. All three of these groups had lower fertility than rural island sedentes. Rindfuss concluded that this indicates urban living *per se* can lower fertility. Another finding was that long-term United States migrants had lower eventual fertility than urban island women. So, assimilation into the cultural 'mainstream' may also be an important factor in fertility.

Migration in the Philippines was studied by Hinday (1978). He found that rural sedente fertility was higher than that of urban migrants, which was higher than that of urban natives. Families of one migrant and one urban native parent had fertility intermediate to the last two groups. Migrants had higher education and married later than rural sedentes. Urbanization had an effect independent of these variables, since the fertility of all groups declined with increased exposure to city living.

A UNESCO report on the effects of rural–urban migration on women in Latin America reviewed recent research in that geographic area (Orlansky & Dubrovsky, 1978). The authors emphasized the bias towards young women in the migration flow. These women often found work as domestics or in semi-skilled occupations, which significantly increased their economic status. Fertility of the migrant women was lower than rural sedentes and sometimes even lower than the urban natives. However, migrant fertility was generally higher than that of the urban born.

Sabagh & Yim (1980) studied the historical and political influences on fertility in Morocco. Moroccan independence occurred in 1956. Pre-1956 migrants to urban areas had the highest fertility of all women, rural sedentes included. Post-1956 migrants had the lowest fertility of all groups. Controlling for age at marriage and socioeconomic differences between groups reduced the magnitude of these findings but did not change the pattern. The authors concluded that greater social mobility and countrywide 'modernization' in the post-independence era may have lowered fertility in those people susceptible to these influences. People willing to migrate appeared to have been the most susceptible.

Three groups of families living in Guatemala City were studied by Bogin & MacVean (1981). Family size (number of children ever born to the parents) was assessed by a questionnaire. Mean family size for the group with two rural-to-urban migrant parents was 4.8 children, for the group with two city-born parents it was 4.5 children, and for the group with one migrant and one city-born parent it was 4.2 children. The first and last means are significantly different. The reason for the difference in fertility between families of different migration status was not clear. Age of the mother, her age at marriage, and length of time in the city were not known. Even so, the fact that the families with one migrant and one city-born parent had the fewest children is unusual. Unlike the study by Martine (1975) the sex of the migrant parent made no difference in mean family size. Bogin & MacVean argue that compared with the other family types, the families with one migrant and one city-born parent had a combination of rural and urban experiences, values, and motivations that may have given them a wider range of adaptive strategies to urban life.

All of the above studies found migrant fertility to be lower than that of rural sedentes. In contrast, the fertility of women migrating from highland areas in the Andes mountains to lowland rural or urban areas increases (Abelson, 1976; Hoff & Abelson, 1976; Foxman, Frerichs & Becht, 1984). One reason for decreased fertility and fecundity at high altitude may be hypoxia. Infant mortality is also higher in the mountains than in lowland rural or urban areas. Undernutrition, infectious disease, cold and other factors besides hypoxia contribute to low fertility and high infant mortality. To compensate for prenatal and postnatal reproductive losses high-altitude peoples may have cultural values and practices that maximize fertility and child survival in their native environment. These values and practices may be carried over to the new lowland environments and result in very high fertility rates in the first generation migrants. Longitudinal studies of fertility adjustment in these migrants are needed to clarify the factors involved in this process.

The question of migrant selection comes up in several of the studies cited above. For instance, people who delay marriage and births should find it easier to migrate. The fact that in much of the world young, unmarried women predominate in the migration flow is evidence for such selection. Lee & Farber (1984) looked at this question of selection in detail for a group of married migrant women. The authors proposed three reasons for the common finding of fertility decline in rural women migrating to the city. These are: (1) selection – migrants are a special group as regards demographics, socioeconomic status, etc., (2) disruption – the act of migration interferes (economically, psychologically, etc.) with fertility goals for a time, and (3) adaptation – migrants assimilate the values of the urban social environment; this may occur even prior to migration and results in lowering of fertility with time. There was no evidence for disruption, because migrant women's fertility did not increase in the years following migration. To examine the effects of selection and adaptation Lee & Farber analysed the data of the Korean World Fertility Survey of 1974 (5000 women aged 20–49, married only once, with at least one live birth). The effect of selection was controlled by matching migrants and rural sedentes for socioeconomic status, education, and stated preferences for family size. The migrants, living in the cities at the time of the survey, had significantly lower fertility than sedentes. Fertility decline persisted as the length of time in the city increased. The results negated selection and supported the adaptation explanation. In this case adaptation was interpreted as a shift from agricultural to wage-earning labour, greater labour force participation for women, and a revision of fertility goals as greater knowledge and experience in the city was obtained.

An important corollary of the Lee & Farber study is that rural-to-urban migration is lowering the overall fertility of Third World populations. Holmes (1976) predicted that rural fertility would increase with time as the low fertility individuals selectively migrated to the city, but the Korean data show that rural-to-urban migration is reducing the fertility rate of that nation as a whole. This is also the case for Latin America (Rogers, 1982), but data for Africa and the rest of Asia are inconclusive at this time. The massive flow of rural-to-urban migrants initially increases the urban population, but as the migrants adapt to the urban environment, fertility decreases. This is exactly what happened in the United States during the last 100 years. In 1940, internal migrants from rural areas still had higher fertility than urban natives, though the youngest migrants were already achieving lower fertility than even the urban-born. By 1960, only the oldest group of rural-born had higher fertility than the urban-born and even they decreased fertility by 20% (Zarate & Zarate, 1975). Because the United States is so highly urbanized, and the urban influence extends into rural areas, fertility has decreased dramatically. If we can apply the experience of the developed nations to the less developed countries we should see similar results in the near future. In other words, since the urban population will soon exceed the rural population in the less developed countries, urban migration should eventually lead to lower total fertility rates in these countries.

Health status and mortality

Migrants are susceptible to the diseases of both their place of origin and their new environments (see Chapter 8). Migration from the countryside to the city represents a drastic change in the physical, social and cultural environment. It was noted long ago by Hippocrates that 'it is changes that are chiefly responsible for disease, especially the violent alteration'. Thus, it is not surprising that many recent studies find that rural-to-urban migrants have higher rates of disease than urban natives (Way, 1976; Velimirovic, 1979; Baker, 1984). It has been noted that urban migrants have increased risks for the development of certain infectious diseases (e.g. tuberculosis), hypertension, coronary heart disease, Type 2 diabetes (diabetes mellitus), gout, and obesity. Other studies show that the health of urban migrants is better than that of the non-migrating rural population (Baker, 1977a; Hollnsteiner & Tacon, 1982). All research shows that since 1950 urban migrants have lower mortality at all ages than rural sedentes (Baker, 1977a; Rogers, 1982).

Placed in an evolutionary and historical context, rural-to-urban migration can be seen to have a changing pattern of impact on human health and mortality. The evolution of human infectious disease is coincident with the development of cities (see section on the extent of urbanization above). The low socioeconomic segments of the urban population were (and still are) also at risk for malnutrition and physically debilitating working conditions, especially for children. As a consequence, until the early part of this century city populations suffered higher infant, childhood and adult mortality than rural populations. For instance, between 1871 and 1880 infant mortality in Sweden was 193/1000 in urban areas and 119/1000 in rural areas. Swedish life expectancy at that time was 43.4 years in the city versus 51.6 years in rural areas. Total life expectancy in England and Wales in 1841 was 40.2 years but only 35 years in London and 24 years in Manchester. Similar statistics hold for Norway and France (Preston, Haines & Panuk, 1981). The census of England and Wales of 1911 shows that childhood mortality prior to age 5 was 20% of live-born children in urban areas and 14% in rural areas. Jorde & Durbize (1986) analysed death certificates for Utah Mormons and found that prior to 1890 mortality rates were generally higher in urban areas than in rural areas. After 1890 the relative frequency of mortality of the two areas was reversed. The 1900 census of the United States shows the same pattern. Expressed as a ratio of actual deaths to statistically expected deaths, urban areas had a ratio of 1.14 and rural areas a ratio of 0.92 (Preston *et al.*, 1981). Even after controlling for the effects of socioeconomic status, working status of the mother, presence or absence of the father, or migrant versus sedente status, the study shows that it was healthier to live in the countryside than in the city. Living conditions for most people in the urban areas of Europe and the United States at that time were like conditions in the urban slums of the less developed countries today.

Today, adaptations to the diseases of urbanization, including genetic and physiological changes, public health programmes, child labour and education laws, and medical treatment, result in generally better health in the urban versus rural areas of the developed and developing world. Rural poverty, malnutrition, and high rates of infectious disease interact synergistically and result in higher rates of morbidity and mortality than in the city (Scrimshaw, 1975; Preston, 1975, 1980; Haines & Avery, 1982). At least this is true for sedente populations. Migrants to the city risk increased exposure to disease by their movement through different ecological zones (e.g. malaria when moving from high-to-low altitude in the tropics), contacting new groups of people, experiencing physical stress (e.g. fatigue, malnutrition) and through the increase in psychological stress related to problems of social adjustment (Prothero, 1977).

Prothero (1977) and Prior (1977) reviewed the general epidemiological importance of migration for human health (see Chapter 8). The effects of urbanization on human health have been treated by Boyden (1970), Harrison & Gibson (1976), and Harrison & Jeffries (1977). The remainder of this section provides a review of the effects of rural-to-urban migration on health, beginning with publications from about 1960 when it appears that interest in this area began.

Gampel *et al.* (1962) compared rural- and urban-living South African Zulu women for blood pressure. The urban-living women were virtually all migrants and had higher rates of hypertension than rural women. However, short-time urban residents had higher rates than long-time residents. The authors found no relationship between diet or obesity and hypertension. They speculated that the emotional stress related to migration may have been a factor promoting increased rates of hypertension in the recent urban migrants. Scotch (1963) found that urban Zulus had higher blood pressure than rural Zulus. Age, sex, obesity, social organization, income, length of urban residence, and use of medical clinics were all related to blood pressure. Those most susceptible to hypertension in the city were individuals who maintained rural cultural practices and did not become integrated into urban life. Migrants from the Easter Islands to cities in Chile showed a marked increase in blood pressure (Cruz-Coke, Etcheverry & Nagel, 1964). Native islanders did not have hypertension at any age. After three to five years, the migrants showed cases of hypertension at all ages. The authors believed that the islanders had a genetic predisposition to hypertension, but required the stress of migration to new environments to bring out the clinical symptoms.

Syme, Hyman & Enterline (1964) cited epidemiological data showing that coronary heart disease had a higher prevalence in urban versus rural areas. From this the authors speculated that urban migrants may be at a greater risk for heart disease. To assess this possibility they studied internal migrants in the United States and found that urban migrants had twice the rate of coronary heart disease compared with rural sedentes. Lifestyle changes associated with urban living seem to have been the culprit, since white-collar workers living in the city or in rural areas had more than twice the risk of coronary heart disease of blue-collar or agricultural workers. An interesting case of reverse migration described by Tyroler & Cassel (1964) supported the association between lifestyle and coronary heart disease. A rural North Carolina county was 'urbanized' by an influx of industry. The coronary heart disease rate for the original sedente population rose following this influx as the native people were forced to alter their values and behaviour.

Rural people migrating to the city of Teheran, Iran, may not face many of the abrupt changes that other urban migrants experience. This is because Teheran is divided into various neighbourhoods based on region of rural origin and length of residence in the city. Even so, migrants, especially women, report higher rates of disease for all ailments than urban natives (Stromberg, Peyman & Dowd, 1974). Nadim, Amini & Malek-Afzadi (1978) compared three groups of people in Iran in order to explain the variation in blood pressure found in that country. The first group was a sample of East Azarbaijani villagers, who were sedentary agriculturalists in a remote region of rural Iran. The second group was a sample of the long-term resident population of Teheran. The third group was a sample of migrants from East Azarbaijan to Teheran. The authors described marked differences between the rural and urban ecosystems, including social roles, cultural values and diet practices. Cigarette smoking, occupation, education and salt intake were recorded and statistically removed as variables influencing blood pressure. It was found that the rural sedentes had lower blood pressure than either of the urban-living groups and that both migrants and long-term residents had similar blood pressure levels. The authors conclude that dietary factors not controlled may have contributed to the results, but social pressures of the Westernized environment of urban Teheran were the primary reasons for higher blood pressure in both urban natives and rural-to-urban migrants.

Benyouseff *et al.* (1974) studied the health effects of rural-to-urban migration in Senegal. They compared Serer peoples living in rural villages with Serer migrants to Dakar. No control for age at migration or length of residence in the city was taken. There were no differences in blood pressure between rural and urban groups. Cholesterol levels were significantly higher in the city, but were still low by World Health Organization standards. Tuberculosis, malaria and schistosomiasis were all more frequent in the rural group. Haematocrit was higher in the urban migrants. This could have been due to less malaria, better nutrition, or both. In sum, urban migrants were healthier than rural sedentes. In a later study Koate (1978) found that hypertension is higher in Dakar than in rural Senegal. Most patients with hypertension were recent migrants from rural villages to the shantytowns at the periphery of the city. The stress of initial urban adaptation and the poverty of the shantytowns may have been responsible for the rise in blood pressure.

The blood pressure of school children from the rural canton of Chateauponsac, France, and of migrant children from the canton to the city of Lyon was measured by Hiernaux & Maquet (1979). At any age between 6–17 years, the systolic blood pressure of the former was 20 mm Hg lower than the latter. The authors suggested three reasons for the difference in blood pressure. These were: (1) lower levels of neuropsychological stimulation in the canton, (2) a faster rate of maturation in the city, and (3) selective migration of families prone to high blood pressure from the canton to the city. Selective migration for blood pressure would have had to have been maintained across generations, and there was little evidence for this. Thus, the first and second reasons for the blood pressure difference were considered more probable as causes than selective migration.

There is evidence of selective migration for blood pressure from the Tecumseh, Michigan, longitudinal study. The blood pressure of females emigrating from the rural town of Tecumseh is higher than that of non-migrating females (Smith & Sing, 1977). Some Polynesian migrants also have higher than average values prior to migration (Prior, 1974). However, it is not known if people with higher than average blood pressure migrate or if the reasons for migration and the contemplation of the move result in higher blood pressure. At this time there is no direct evidence that migration is selective for blood pressure.

There is evidence that lifestyle changes occurring before or after migration influence blood pressure. For instance, Weitz (1982) found that the systolic blood pressure of migrants from rural villages in highland Tibet to the city of Kathmandu was higher than that of the village sedentes and migrants to other high-altitude villages. The higher blood pressure of the urban migrants was not related to altitude or morphology (e.g.

fatness), but rather to Western influences on activity patterns, diet and lifestyle.

Returning to Africa, Poulter *et al.* (1984) found that Kenyans of the Luo Tribe had higher blood pressure in urban areas than in rural villages and that the blood pressure of urban migrants rose after their move to the city. Increases in body weight and urinary electrolytes (probably due to salt consumption) were related to the blood pressure rise in men, while only urinary electrolytes were associated with the blood pressure rise in women. These factors correlated with changes in diet for urban migrants, shifting from maize, green vegetables and occasional meat to more meat, salt and fewer vegetables.

Epidemiological surveys have been carried out among South Pacific islanders from New Zealand (Maoris), Tonga, Cook Island, Tokelau, Samoans and Tokelau migrants to New Zealand since 1962 (Prior *et al.*, 1974; Finau, Prior & Evans, 1982; Baker, 1984). Information on the health status of migrants to the cities of these islands is available. The major findings are that after urban migration, obesity increases, serum lipids show a decrease in the high density lipoprotein fraction, blood pressure may increase (but not in all groups), coronary heart disease and Type 2 diabetes incidence increases, and mortality at all ages is generally higher. The changes with migration are not uniform. Urban-living Samoans become some of the fattest people in the world (Pawson & Janes, 1981), but Tokelauan migrants do not become so obese (Baker, 1984). Finau *et al.* (1982) cited changes in lifestyle following urban migration as the reason for the deterioration of health in the cities. The consumption of new foods, often processed with more salt, more refined carbohydrates and less fibre, plus the increased use of tobacco and alcohol, were listed as causes of higher urban morbidity and mortality. The shift from agriculture and fishing to sedentary occupations also contributed to the problem. Finally, the breakdown of traditional culture added a psychosocial stress that may have brought on or exacerbated illness. An issue not specifically covered by these studies is how low socioeconomic status and racial prejudice might negatively influence the native migrants to Europeanized cities in New Zealand and elsewhere. The well-known difference in morbidity and mortality rates for Whites and Blacks in the United States suggests that racism may be a significant factor in the South Pacific as well.

South Pacific migrants to urban areas are eventually able to adapt to the change in lifestyle (see also Chapter 7). Baker and his colleagues use a biosocial approach to study changes in blood pressure due to migration and modernization in Samoa and Hawaii. McGarvey & Baker (1979) compared Samoans from traditional rural areas and more modern areas in Samoa to Samoan migrants to Hawaii. Sedente blood pressure was higher in modern areas. Among migrants, those from the traditional areas of Samoa had higher blood pressure than their sedente counterparts. Migrants from modernized areas of Samoa had blood pressures equal to that of their non-migrant peers. When all the migrants were compared, those most in contact with modern life had blood pressures lower than migrants with intermediate contact. This finding was repeated in a study by Hanna & Baker (1979). Samoan migrants living in Honolulu had lower blood pressure than migrants living in rural, traditional areas of Hawaii. In both studies, blood pressure was also positively correlated with fatness. Though fatness tended to increase with migration and age, the relationships were not linear. Rather, fatness continued to increase with greater time in the urban environment, while blood pressure at first rose with fatness and then fell with further time in the city. Bindon & Baker (1985) examined the relationship of modernization, migration and obesity among Samoan sedentes and migrants to Hawaii. They found that migration, *per se*, did not affect the frequency of obesity, even when adjustments were made for length of time in Hawaii or the fraction of a subject's lifetime spent in Hawaii. Living in a modern society did lead to a rapid increase in the frequency of obesity for Samoan sedentes in American Samoa and migrants to Hawaii. American Samoan women developed the greatest incidence of obesity of all groups. These women retained their traditional rural occupational activities, but lived in households that depended on wage labour for subsistence. Thus the women were excluded from significant cash-earning roles in these households. Migrant men developed their highest incidence of obesity when they lived in the total cash economy of Hawaii. Education was negatively correlated with obesity, especially for young women. As was the case with blood pressure, acculturation to Western society via education and other means may eventually lead to a return of more desirable levels of fatness and health.

The risk for Type 2 diabetes in the city has been analysed by Zimmet (1982) in an epidemiological review of this disease. He included data from the South Pacific Island studies just mentioned and other studies. In Western Samoa the prevalence rate for diabetes was 2.7% for rural sedentes and 7.0% for urban migrants. In Fiji the corresponding rates were 1.8% versus 6.9% for Melanesians. In the Cook Islands the rates are 2.4% versus 5.7%. Finally, in India the rates were 1.2% versus 2%. All these rates were low compared with those for the Nauru of Micronesia, 30.3%, and the Pima Indians of the American Southwest, 34.1%. These last two groups were still rural-living, and Zimmet believes that their high incidence for Type 2 diabetes was due to genetic factors more than lifestyle variables.

Diabetes mortality among Chinese migrants to New York City was studied by Gerber (1984). The majority of the migrants came from Kwangtung Province in south China. They were a mixture of rural villagers and city dwellers, though very few came from cities like New York. Chinese, in general, had a higher proportion of deaths due to diabetes than Whites or non-Whites living in New York. China-born Chinese had a slightly higher rate of death than Chinese born in the United States. Finally, the proportion of deaths from diabetes increased with age in all groups, but especially in the China-born Chinese. Moreover, the recent migrants seemed to suffer more than migrants with more time spent in the United States. The reasons for this are not known. In another study, Gerber & Madhaven (1980) have shown that Chinese migrants to the United States have a lower proportion of deaths from heart disease than either Chinese born in the United States or Whites living in New York. This suggests that assimilation to United States urban culture decreased the risk of coronary heart disease while at the same time it increased the risk for diabetes death. In that both coronary heart disease and diabetes have lifestyle concomitants, and often the same ones such as obesity and sedentary behaviour, these findings are intriguing, but not explicable at present.

A series of minor diseases also show an increase from rural to urban populations. These minor diseases include malocclusion, dental caries, myopia and chronic allergy (Corruccini & Kaul, 1983). Many of these conditions such as malocclusion and myopia are traditionally considered to be genetic and hereditary. However, Lasker (1945) found that several dental characteristics of Chinese migrants to the United States were not genetically controlled: instead, they were influenced by the change in environment. For instance, dental caries occurred at a much higher incidence in Chinese living in the United States than in China, and the rate of caries increased with total time spent in the United States. Crowding of teeth was another characteristic that increased with migration to the United States. More recent research supports the idea that these minor diseases and conditions are linked to urbanization and Western lifestyle (Corruccini & Lee, 1984; Corruccini, 1984; and see Corruccini & Kaul, 1983, for a general review). In particular, diet patterns, including a dependence on refined or highly processed foods, correlate with malocclusion, and reading and close-study school work may lead to myopia. Not all urban migrants develop all or any of these

116   B. Bogin

problems, but studies of internal migrants in India, Africa, and the Americas, and international migrants from rural China to urban England do show many of them.

What general conclusions can be made from all of these studies? The first conclusion is that most of the research on health and migration is retrospective and essentially limited to census-type data. These studies demonstrate that migration and health changes are correlated, but they cannot be used to determine the true cause and effect relationships.

A second conclusion is that the typical pattern of migration and health status may be represented by a 'U'-shaped curve. Gullahorn & Gullahorn (1963) described the U-curve response in terms of emotional adjustment to voluntary migration. First there is an emotional 'high' of excitement prior to the move, then there is a period of evaluation of the new surroundings, then a period of depression as problems mount and solutions seem unattainable, and finally there is a return to emotional balance as the migrant adapts to his new surroundings. The U-curve hypothesis predicts that rural-to-urban migrants with an intermediate degree of contact with the urban environment should be under the most stress. The few studies of a prospective, longitudinal, and/or multidisciplinary nature, such as those by Baker and his colleagues reviewed above, do show that health usually deteriorates following migration but then improves as the length of residence in the new urban environment increases. There are exceptions for specific diseases, such as coronary heart disease, obesity, and some cancers which increase in incidence with time following migration.

Support for the U-curve hypothesis comes from a study of Filipino migrants living on Oahu, Hawaii (Brown, 1982). General stress levels were evaluated by measuring 24-hour excretion rates of urinary catecholamines. Excretion rates increased in direct response to levels of physiological stress. Migrants with either low or high degrees of contact with urban Hawaiian society had lower levels of catecholamine excretion than migrants with intermediate degrees of contact. Increased stress may predispose an individual to both infectious and non-infectious disease (Dubos, 1965; Carlstram and Levi, 1971). Thus, migrants attempting to adapt and acculturate themselves to a new urban environment should experience an increase in morbidity, and they generally do. Eventually these migrants seem to adapt, in that their health status improves, and they live longer than their rural non-migrating peers. Their psychosocial adjustment no doubt lowers stress levels, but the greater availability of public health facilities, medical services, social services and education, and the more reliable supply and greater variety of food in the city correlate with the eventual improvement in health and longevity that most migrants can expect to enjoy.

Effects of migration on the remaining population

The rural-to-urban migration literature is replete with studies of the biological effects of migration on migrants after they have reached their destination. Some studies also consider the effects of immigration on the recipient population. Few, studies examine the biological changes that occur in the non-migrating rural position.

The social and economic literature on migration may serve as a starting point. Rural-to-urban migrants are generally better educated, of higher socioeconomic status, younger, and less traditional in cultural values than rural sedentes (Illsley *et al.*, 1963; Kennedy, 1973; Butterworth, 1977; Dubisch, 1977; Jenkins, 1977; Butterworth & Chance, 1981). It is well known that these social characteristics correlate with physical size, fertility, health, and mortality. Because of this social selection, rural sedentes tend to be shorter and lighter (Mascie-Taylor, 1984), more fertile (Zarate & Zarate, 1975), suffer more from the diseases of poverty – malnutrition, preventable infectious diseases and parasitic diseases (Baker, 1977*a*), and experience higher mortality at all ages compared with urban migrants (Rogers, 1982), but only after the migrants arrive and adapt to the lifestyle of the city. As was shown above in the section on growth and development, there are few biological differences between eventual migrants and sedentes before the act of migration takes place.

Selective emigration for age and sex has effects on rural demography. The urban flow of the young, especially young women, results in overall rural depopulation despite the higher fertility of the sedentes. This has happened in the developed nations, where natural increase in the cities is greater than in the rural areas. It is now occurring in the less developed countries, where by the year 2000 the natural increase of urban populations will be greater than that of rural populations (United Nations, 1980; Rogers, 1982). An extreme case of this demographic effect occurred in Ireland (Kennedy, 1973). The natural rate of increase in the rural counties during the last century would have produced a total national population of over 20 million people by 1970. Massive rural emigration to the cities of England and the United States, however, resulted in a total national population of just over two million.

The social selection of urban migrants does have a limit in terms of its effect on the biology of rural sedentes. The rural non-migrants of the developed nations have not become a homogeneous group for stature, fertility or total health characteristics. Pronounced variation at the genetic and phenotypic levels are still measurable – even in rural Ireland (Clegg, 1979; Relethford, Lees & Crawford, 1980; Boyce, 1984). Much of this variation is caused by the movement of people between rural areas, a topic that deserves its own treatment. It is also due to the spread of modernization, urban values and some aspects of urban lifestyle into the countryside. As this occurs and the rural environment changes, especially in terms of diet and activity patterns for children, human biology changes as a result of our species' capacity for developmental plasticity. This process of change is occurring today in the less developed countries, as evidenced by the continuation of secular trends in growth and maturation (Roche, 1979). In the most developed nations, such as the United States and England, secular trends appear to have ceased and there are no longer any rural-to-urban differences in stature, fertility and mortality when socioeconomic differences are controlled. In other words, the urban poor and the rural poor suffer equally.

In sum, rural sedentes can be differentiated both socially and biologically from urban-living migrants. By our current criteria for biological 'well-being', rural sedentes in all but the most developed nations are at a disadvantage. However, variability within each group is so great that it is not possible to characterize biologically the typical rural sedente nor the 'average' urban migrant. Indeed, social variation accounts for most of the biological differences. Throughout history, the 'perversity of urbanization' (Keyfitz, 1982) has drained away the wealthiest, best educated, youngest and healthiest from the countryside. Eventually, though, the country folk usually adopt many of the city ways and enjoy the benefits and liabilities of urbanization without experiencing the trauma of migration.

Conclusion: migration and adaptation to the city

Fig. 5.1 is a summary of the process and the biological consequences of urban migration discussed in this review. This figure applies to the migration process now taking place in the less developed countries and the process that took place in the more developed countries between 1750 and 1940. The extreme left column begins with the rural-living individual. Moving down this column, the next box characterizes the social profile of likely sedentes. Some major rural environmental variables that influence human biology are listed in the next box. The last box in the first column lists the biological correlates of sedentism. The social profile of likely rural-to-urban migrants is given in the upper box of the second column.

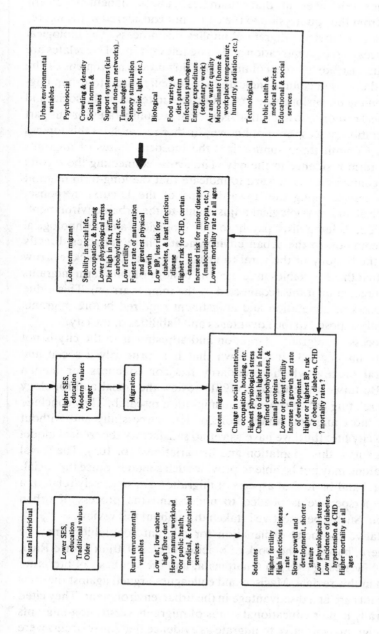

Fig. 5.1. Flow diagram summarizing the major biosocial correlates and consequences of rural-to-urban migration in the least developed countries today and in the more developed nations prior to 1940. Abbreviations used are: BP, blood pressure; CHD, coronary heart disease.

The complex series of events that culminates in the permanent movement of people from the countryside to the city is not considered in the figure. Instead the box labeled 'Migration' implies this process. The biological consequences of recent migration and some relevant social correlates are given in the last box of this column. The downward movement of the arrows in the second column indicate that, initially, rural-to-urban migrants are on the downslope side of the U-curve in terms of the social and physiological response to migration. Note that mortality rates for this part of the migration process are not known; further research on this topic is necessary. Column three summarizes the biosocial status of migrants after long-term residence in the city. The arrow connecting the second and third column curves upward to indicate that the long-term migrant eventually moves along the upward slope of the U-curve response, making social and physiological adjustments to the urban environment. The far right column lists psychosocial, biological and technological variables common to the urban environment. These variables directly influence the biology of the rural-to-urban migrant. The thicker arrow indicates that these variables influence the biology of long-term migrants more than recent migrants (connected by the thinner arrow). This is due to the process of adaptation and adjustment required before migrants become fully exposed to the advantages and liabilities of the city.

The process of migrant adaptation and adjustment to the city is not completely understood. It is known that it is primarily a social and behavioural process which secondarily has consequences for human biology. Butterworth & Chance (1981) review the social anthropology literature relating to Latin American migration research. They conclude that '. . . most migrants adapt more or less successfully and without trauma to city life, [but] we have as yet no satisfactory theoretical model that can explain this adaptation and its variations' (p. 103). The social research alone may not be able to provide such a model. Since the social, behavioural and biological aspects of migration are so closely related, a biosocial perspective is needed to understand the process of urban adaptation. Such a perspective is taken throughout this review.

A tentative biosocial explanation for migrant adaptability may be offered here, based on the work of Macisco et al. (1970) in Puerto Rico, Bogin & MacVean (1981) in Guatemala, and the synthesis of literature discussed in this review. Macisco and colleagues argued against the view that migrants are at a disadvantage in the urban environment. They cited the generally higher educational status of migrants versus non-migrants and the migrants' initiative to migrate as evidence that the migrants were able and willing to make the necessary adjustments to urban life. These researchers speculated that migrants may have possessed a greater repertoire of adaptive strategies to the stress of the city, because of their combined rural and urban experience, than even those born in the city. Bogin & MacVean found that many migrants to Guatemala City married urban natives. These 'mixed' families appeared to adapt more successfully than either families with two migrant parents or two city-born parents, the mixed parent families had the fewer children, taller children, and a socioeconomic status at least equal to the city-born families. These mixed families may have had an even greater repertoire of urban and rural adaptive strategies than the Puerto Rican migrants, which may have accounted for their lower fertility and greater growth of their children. Thus, personal initiative, higher-than-average socioeconomic status, and ability to draw upon rural and urban experiences are one set of factors that may explain the adaptability of rural-to-urban migrants.

The review of literature presented in this chapter, and summarized in Fig. 5.1, shows that rural-to-urban migration results in a series of biological changes in the growth and development, fertility, morbidity and mortality of migrants and their descendants. The biological changes are the result of adaptation and adjustment to the urban biosocial environment. The process of adaptation requires behavioural and social flexibility, which are common characteristics shared by urban migrants. Migrants are not a random sample of the rural population: they are of higher socioeconomic status and are younger than rural sedentes. This attests to their social selection. The biological changes that occur in these people and their descendants following migration to the city are a result of the social selection, their social and behavioural flexibility, and the fundamental plasticity in human biology to respond to environmental change. Rural-to-urban migrants do adapt socially and biologically. The potential nightmare of a world of overcrowded, impoverished cities can be tempered by the realization of this fact.

References

Abelson, A. E. (1976). Altitude and fertility. *Human Biology*, **48**, 83–91.
Ammon, O. (1899). Zur Anthropologie der Badener. Jena, cited in Boas, F. (1922).
Armelagos, G. J. & Dewey, J. R. (1970). Evolutionary response to human infectious diseases. *Bioscience*, **20**, 271–5.
Baker, P. (1977a). Problems and strategies. In *Human Population Problems in the Biosphere: Some Research Strategies and Designs*, MAB Technical Notes 3, ed. P. T. Baker, pp. 11–32. Paris: UNESCO.
Baker, P. (1977b). Biological and social aspects of migration of the Andes population. *Archivos de Biologia Andina*, **7**, 63–82.

Baker, P. (1984). Migration, genetics, and the degenerative diseases of South Pacific Islanders. In *Migration and Mobility*, ed. A. J. Boyce, pp. 209–39. London: Taylor and Francis.

Benedict, B. (1972). Social regulation of fertility. In *The Structure of Human Populations*, ed. G. A. Harrison & A. J. Boyce, pp. 73–89. Oxford: Clarendon Press.

Benyousseff, A., Cutler, A., Levine, P. *et al.* (1974). Health effects of rural–urban migration in developing countries – Senegal. *Social Science and Medicine*, 8, 243–54.

Bindon, J. R. & Baker, P. T. (1985). Modernization, migration and obesity among Samoan adults. *Annals of Human Biology*, 12, 67–76.

Boas, F. (1912). Changes in the bodily form of descendants of immigrants. *American Anthropologist*, 14, 530–63.

Boas, F. (1922). Report on an anthropometric investigation of the population of the United States. *Journal of the American Statistical Association*, 18, 181–209.

Boas, F., (ed). (1940). *Race, language and culture*. New York: Free Press.

Bogin, B. & MacVean, R. B. (1978). Growth in height and weight of urban Guatemalan primary school children of high and low socioeconomic class. *Human Biology*, 50, 477–88.

Bogin, B. & MacVean, R. B. (1981). Biosocial effects of urban migration on the development of families and children in Guatemala. *American Journal of Public Health*, 71, 1373–7.

Bogin, B. & MacVean, R. B. (1984). Growth status of non-agrarian, semi-urban living Indians in Guatemala. *Human Biology*, 56, 527–38.

Boyce, A. J. (ed.) (1984). *Migration and Mobility*. London: Taylor and Francis.

Boyden, S. V. (ed.) (1970). *The Impact of Civilization on the Biology of Man*. Toronto: University of Toronto Press.

Boyden, S. V. (1972). Ecology in relation to urban population structure. In *The Structure of Human Populations*, ed. G. A. Harrison & A. J. Boyce, pp. 411–41. Oxford: Clarendon Press.

Briggs, A. (1983). The environment of the city. In *How Humans Adapt: A Biocultural Odyssey*, ed. D. J. Ortner, pp. 371–87. Washington, DC: Smithsonian Institution.

Brown, D. E. (1982). Physiological stress and culture change in a group of Filipino-Americans: a preliminary investigation. *Annals of Human Biology*, 9, 553–63.

Butterworth, D. (1977). Selectivity of out-migration from a Mixtec community. *Urban Anthropology*, 6, 129–39.

Butterworth, D. & Chance, J. K. (1981). *Latin American Urbanization*. Cambridge: Cambridge University Press.

Carlstram, G. & Levi, L. (1971). *Urban Conglomerates as Psychosocial Human Stressors*. Stockholm: Royal Ministry for Foreign Affairs.

Chaundhury, R. H. (1978). Female status and fertility behavior in a metropolitan urban area of Bangladesh. *Population Studies*, 32, 261–73.

Childe, V. G. (1942). *What Happened in History*. London: Penguin Books.

Clegg, E. J. (1979). ABO and Rh blood groups in Outer Hebrides. *Annals of Human Biology*, 6, 457–70.

Clegg, E. J. & Garlick, J. P. (eds.) (1981). *Disease and Urbanization.* Symposia of the Society for the Study of Human Biology, vol. **20**. Atlantic Highlands, NJ: Humanities Press.

Cockburn, T. A. (ed.) (1967). *Infectious Diseases: Their Evolution and Eradication.* Springfield, Ill.: Charles C. Thomas.

Corruccini, R. S. (1984). An epidemiologic transition in dental occlusion in world populations. *American Journal of Orthodontics*, **86**, 419–26.

Corruccini, R. S. & Kaul, S. S. (1983). The epidemiological transition and the anthropology of minor and chronic non-infectious diseases. *Medical Anthropology*, **7**, 36–50.

Corruccini, R. S. & Lee, G. T. R. (1984). Occlusal variation in Chinese immigrants to the United Kingdom and their offspring. *Archives of Oral Biology*, **29**, 779–82.

Cruz-Coke, R., Etcheverry, R. & Nagel, R. (1964). Influence of migration on blood pressure of Easter Islanders. *The Lancet*, **28**, 697–9.

Davies, C. T. M., Mbelwa, D. & Dore, C. (1974). Physical growth and development of urban and rural East African children, aged 7–16. *Annals of Human Biology*, **1**, 257–68.

De Villiers, H. (1971). A study of morphological variables in urban and rural Venda male populations. In *Human Biology of Environmental Change*, ed. D. J. M. Vorster. London: International Biological Program.

Dubisch, J. (1977). The city as resource: migration from a Greek island village. *Urban Anthropology*, **6**, 65–81.

Dubos, R. (1965). *Man Adapting.* New Haven: Yale University Press.

Finau, S. A., Prior, I. A. & Evans, J. G. (1982). Aging in the South Pacific: physical changes with urbanization. *Social Science and Medicine*, **16**, 1539–49.

Foxman, B., Frerichs, R. R. & Becht, J. N. (1984). Health status of migrants. *Human Biology*, **56**, 129–41.

Freedman, R., Whelpton, P. K. & Campbell, A. A. (1959). *Family Planning, Sterility and Population Growth.* New York: McGraw-Hill.

Frisancho, A. R., Klayman, J. E. & Matos, J. (1976). Symbiotic relationship of high fertility, high childhood mortality and socio-economic status in an urban Peruvian population. *Human Biology*, **48**, 101–11.

Gampel, B., Slome, C., Scotch, N. & Abramson, J. H. (1962). Urbanization and hypertension among Zulu adults. *Journal of Chronic Disease*, **15**, 67–70.

Gerber, L. M. (1984). Diabetes mortality among Chinese migrants to New York City. *Human Biology*, **56**, 449–58.

Gerber, L. M. & Madhaven, S. (1980). Epidemiology of coronary heart disease in migrant Chinese populations. *Medical Anthropology*, **4**, 307–20.

Goldscheider, C. (1971). *Population, Modernization and Social Change.* Boston: Little, Brown.

Goldstein, M. S. (1943). *Demographic and bodily changes in descendants of Mexican immigrants.* Austin: Institute of Latin American Studies.

Goldstein, S. (1973). Interrelations between migration and fertility in Thailand. *Demography*, **10**, 225–58.

Graham, G. G., MacLean, W. C., Jr, Kallman, C. H., Rabold, J. & Mellits, E. D. (1979). Growth standards for poor urban children in nutrition studies. *American Journal of Clinical Nutrition*, **32**, 703–10.

124   B. Bogin

Gullahorn, J. T. & Gullahorn, J. E. (1963). An extension of the U-curve hypothesis. *Journal of Social Issues*, **19**, 33–47.
Haines, M. R. & Avery, R. C. (1982). Differential infant and child mortality in Costa Rica: 1968–1973. *Population Studies*, **36**, 1–13.
Hanna, J. M. & Baker, P. T. (1979). Biocultural correlates to the blood pressure of Samoan migrants to Hawaii. *Human Biology*, **51**, 461–97.
Harrison, G. A. & Gibson, J. B. (eds) (1976). *Man in Urban Environments.* Oxford: Oxford University Press.
Harrison, G. A. & Jeffries, D. J. (1977). Human biology in urban environments: a review of research strategies. In *Human Population Problems in the Biosphere: some Research Strategies and Designs*, MAB Technical Notes 3, ed. P. T. Baker, pp. 65–82. Paris: UNESCO.
Hauser, P. M. & Schnore, L. F. (1965). *The Study of Urbanization*. New York: Wiley.
Hiernaux, J. & Maquet, D. (1979). La population du Canton rural de Chateauponsac (Haute-Vienne, France) et ses émigrés en ville, II. *Bulletin et Mémoires de la Société d'Anthropologie, Paris*, **6**, 181–9.
Hinday, V. A. (1978). Migration, urbanization and fertility in the Philippines. *International Migration Review*, **12**, 370–85.
Hoff, C. J. & Abelson, A. E. (1976). Fertility. In *Man in the Andes: a Multidisciplinary Study of High Altitude Quecha*, ed. P. T. Baker & M. A. Little, pp. 128–46. Stroudsburg, Penn: Dowden, Hutchinson and Ross.
Hollnsteiner, M. R. & Tacon, P. (1982). *Urban migration in developing countries: consequences for families and children.* Paper presented at the annual meetings of the American Association for the Advancement of Science, January 4, Washington, DC.
Holmes, D. N. (1976). Migration and fertility: introduction. In *The Dynamics of Migration: Internal Migration and Fertility.* Occasional Monograph Series 1(5). Washington DC: Interdisciplinary Communications Program, Smithsonian Institution.
Hulse, F. (1969). Migration and cultural selection in human genetics. In *The Anthropologist*, ed. P. C. Biswas, pp. 1–21. Delhi, India: University of Delhi.
Hutchinson, B. (1961). Fertility, social mobility and urban migration in Brazil. *Population Studies*, **14**, 182–9.
Illsley, R., Finlayson, A. & Thompson, B. (1963). The motivation and characteristics of internal migrants: a socio-medical study of young migrants in Scotland. *Milbank Memorial Fund Quarterly*, **41**, 115–44 and **41**, 217–48.
Jenkins, J. C. (1977). Push/pull in recent Mexican to the US migration. *International Migration Review*, **11**, 178–89.
Johnson, T. O. (1970). Height and weight patterns of an urban African population sample in Nigeria. *Tropical Geography and Medicine*, **22**, 65–76.
Johnston, F. E., Borden, M. & MacVean, R. B. (1973). Height, weight, and their growth velocities in Guatemalan private school children of high socioeconomic class. *Human Biology*, **45**, 627–41.
Johnston, F. E., Low, S. M., Baessa, Y. de & MacVean, R. B. (1985). Growth status of disadvantaged urban Guatemalan children of a resettled community. *American Journal of Physical Anthropology*, **68**, 215–24.
Jorde, L. B. & Durbize, P. (1986). Opportunity for natural selection in Utah Mormons. *Human Biology*, **58**, 97–114.

Kaplan, B. (1954). Environment and human plasticity. *American Anthropologist*, **56**, 780–99.

Kennedy, R. E. Jr (1973). *The Irish: Emigration, Marriage, and Fertility*. Berkeley: University of California Press.

Ketkar, S. L. (1979). Determinants of fertility in a developing society: the case of Sierra Leone. *Population Studies*, **23**, 479–88.

Keyfitz, N. (1982). Development and the elimination of poverty. *Economic Development and Culture Change*, **30**, 649–70.

Koate, P. (1978). L'hypertension arterielle en Afrique noire. *Bulletin of the World Health Organization*, **56**, 841–8.

Kobyliansky, E. & Arensburg, B. (1974). Changes in morphology of human populations due to migrations and selection. *Annals of Human Biology*, **4**, 57–71.

Kumudini, D. (1965). The effect of education on fertility. *Proceedings of the World Population Conference, Belgrade*, **4**, 146–9.

Lasker, G. W. (1945). Observations on the teeth of Chinese born and reared in China and America (including data on Peking prisoners collected by Liang Ssu-yung). *American Journal of Physical Anthropology*, **3**, 129–50.

Lasker, G. W. (1946). Migration and physical differentiation. *American Journal of Physical Anthropology*, **4**, 273–300.

Lasker, G. W. (1952). Environmental growth factors and selective migration. *Human Biology*, **24**, 262–89.

Lasker, G. W. (1954). The question of physical selection of Mexican migrants to the United States of America. *Human Biology*, **26**, 52–8.

Lasker, G. W. & Evans, F. G. (1961). Age, environment and migration: further anthropometric findings on migrant and non-migrant Mexicans. *American Journal of Physical Anthropology*, **19**, 203–11.

Lee, B. S. & Farber, S. C. (1984). Fertility adaptation by rural–urban migrants in developing countries: the case of Korea. *Population Studies*, **38**, 141–55.

Liberty, M., Hughey, D. & Scaglion, R. (1976). Rural and urban Omaha fertility. *Human Biology*, **48**, 59–71.

Liberty, M., Scaglion, R. & Hughey, D. (1976) Rural and urban Seminole fertility. *Human Biology*, **48**, 741–55.

Lipton, M. (1977). *Why Poor People Stay Poor: Urban Bias in World Development*. Cambridge, Mass.: Harvard University Press.

Livi, R. (1896). *Antropometrica Militare*. Rome, cited in Boas, F. (1922).

Macbeth, H. M. (1984). The study of biological selectivity in migrants. In *Migration and Mobility*, ed. A. J. Boyce, pp. 195–207. London: Taylor and Francis.

Macisco, J. J., Bouvier, L. F. & Renzi, M. J. (1969). Migration status, education and fertility in Puerto Rico, 1960. *Milbank Memorial Fund Quarterly*, **47**, 167–87.

Macisco, J. J., Bouvier, L. F. & Weller, R. H. (1970). The effect of labour force participation on the relation between migration status and fertility in San Juan, Puerto Rico. *Milbank Memorial Fund Quarterly*, **48**, 51–70.

Malina, R. M., Bushang, P. H., Aronson, W. L. & Selby, H. (1982). Childhood growth status of eventual migrants and sedentes in a rural Zapotec community in the valley of Oaxaca, Mexico. *Human Biology*, **54**, 709–16.

Malina, R. M., Himes, J. H., Stepick, C. D., Lopez, F. G. & Buschang, P. H.

126   B. Bogin

(1981). Growth of rural and urban children in the Valley of Oaxaca, Mexico. *American Journal of Physical Anthropology*, **55**, 269–80.

Martin, W. J. (1949). *The Physique of Young Adult Males*. Medical Research Council Memorandum **20**. London: HMSO.

Martine, G. (1975). Migrant fertility adjustment and urban growth in Latin America. *International Migration Review*, **9**, 177–91.

Mascie-Taylor, C. G. N. (1984). The interaction between geographical and social mobility. In *Migration and Mobility*, ed. A. J. Boyce, pp. 161–78. London: Taylor and Francis.

McGarvey, S. T. & Baker, P. T. (1979). The effects of modernization and migration on Samoan blood pressures. *Human Biology*, **51**, 461–79.

McNeill, W. H. (1976). *Plagues and Peoples*. New York: Doubleday.

McNeill, W. H. (1979). Historical patterns of migration. *Current Anthropology*, **20**, 95–102.

Meredith, H. V. (1979). Comparative findings on body size of children and youths living at urban centers and in rural areas. *Growth*, **43**, 95–104.

Michelson, W. H. (1970). *Man and his urban Environment: A Sociological Approach*. Reading, Mass.: Addison-Wesley.

Morley, D. C., Woodland, M., Martin, W. J. & Allen, I. (1968). Heights and weights of West African children from birth to age of five. *West African Medical Journal*, **17**, 8–13.

Mumford, L. (1956). The natural history of urbanization. In *Man's role in changing the face of the Earth*, ed. W. E. Thomas. Chicago: University of Chicago Press. Reprinted in *The Ecology of Man: an Ecosystems Approach*, ed. R. L. Smith, pp. 140–52. New York: Harper & Row.

Myers, G. C. & Morris, E. W. (1966). Migration and fertility in Puerto Rico. *Population Studies*, **20**, 85–96.

Nadim, A., Amini H. & Malek-Afzadi, H. (1978). Blood pressure and rural–urban migration in Iran. *International Journal of Epidemiology*, **7**, 131–8.

O'Dell, M. E. (1984). The children of conquest in the new age: ethnicity and change among the highland Maya. *Central Issues in Anthropology*, **5**, 1–15.

Oppenheim, A. L. (1974). *Ancient Mesopotamia*. Chicago: University of Chicago Press.

Orlansky, D. & Dubrovsky, S. (1978). *The Effects of Rural–Urban Migration on Women's Role and Status in Latin America*. Reports and Papers in the Social Sciences **41**. Paris: UNESCO.

Panek, S. & Piasecki, M. (1971). Nowa Huta: integration of the population of in the light of anthropometric data. *Materialyi i Prace Anthropologiczne*, **80**, 1–249 (in Polish with English summary).

Pawson, I. G. & Janes, C. (1981). Massive obesity in a migrant Samoan population. *American Journal of Public Health*, **71**, 508–13.

Poulter, N., Khaw, K. T., Hopwood, B. E. C., Mugambi, M., Peart, W. S. & Sever, P. S. (1984). Salt and blood pressure in various populations. *Journal of Cardiovascular Pharmacology*, **6**, supp 1, s197–s203.

Preston, S. H. (1975). The changing relation between mortality and level of economic development. *Population Studies*, **29**, 231–48.

Preston, S. H. (1980). Causes and consequences of mortality declines in less developed countries during the twentieth century. In *Demographic and*

Economic Change in Developing Countries, ed. R. A. Easterlin, pp. 289–360. Chicago: University of Chicago Press.

Preston, S. H., Haines, M. R. & Panuk, E. (1981). Effect of industrialization and urbanization on mortality in developed countries. In *International Population Conference, Manila, 1981*, Proceedings vol. 2, pp. 233–53. Liège: International Union for the Scientific Study of Population.

Prior, I. A. M. (1974). Cardiovascular epidemiology in New Zealand and the Pacific. *New Zealand Medical Journal*, **80**, 345–52.

Prior, I. A. M. (1977). Migration and physical illness. *Advances in Psychosomatic Medicine*, **9**, 105–31.

Prior, I. A. M., Stanhope, J. M., Evans, J. G. & Salmond, C. E. (1974). The Tokelau Island migrant study. *International Journal of Epidemiology*, **3**, 225–32.

Prothero, R. M. (1977). Disease and mobility: a neglected factor in epidemiology. *International Journal of Epidemiology*, **6**, 259–67.

Ravenstein, E. G. (1885). The laws of migration. *Royal Statistical Society*, **48**, part 2, 167–277.

Relethford, J. H., Lees, F. C. & Crawford, M. H. (1980). Population structure and anthropometric variation in rural western Ireland: migration and biological differentiation. *Annals of Human Biology*, **7**, 411–28.

Rindfuss, R. R. (1976). Fertility and migration: the case of Puerto Rico. *International Migration Review*, **10**, 191–203.

Robinson, W. C. (1963). Urbanization and fertility: the non-western experiences. *Milbank Memorial Fund Quarterly*, **4**, 291–308.

Roche, A. F. (ed.) (1979). Secular trends in human growth, maturation, and development. *Monographs of the Society for Research in Child Development*, **44(179)**.

Rogers, A. (1982). Sources of urban population growth and urbanization, 1950–2000: a demographic accounting. *Economic Development and Culture Change*, **30**, 483–506.

Rogers, A. & Williamson, J. C. (1982). Migration, urbanization, and third world development: an overview. *Economic Development and Cultural Change*, **30**, 463–82.

Sabagh, G. & Yim, S. B. (1980). The relationship between migration and fertility in a historical context: the case of Morocco in the 1960's. *International Migration Review*, **14**, 525–38.

Scholl, T. O., O'Dell, M. E. & Johnston, F. E. (1976). Biological correlates of modernization in a Guatemala highland municipio. *Annals of Human Biology*, **3**, 23–32.

Scotch, N. A. (1963). Sociocultural factors in the epidemiology of Zulu hypertension. *American Journal of Public Health*, **53**, 1205–12.

Scrimshaw, S. C. M. (1975). Infant mortality and behavior in the regulation of family size. *Population and Development Review*, **4**, 383–403.

Shapiro, H. L. (1939). *Migration and Environment*. Oxford: Oxford University Press.

Smith, D. G. & Sing, C. F. (1977). Genetic–environmental interactions in the variation of blood pressure in Tecumseh, Michigan. *Journal of Chronic Disease*, **30**, 781–91.

128   B. Bogin

Smith, M. T. (1984). The effects of migration on sampling in genetical surveys. In *Migration and Mobility*, ed. A. J. Boyce, pp. 97–110. London: Taylor and Francis.

Steegmann, A. T. (1985). 18th century British military stature: growth cessation, selective recruiting, secular trends, nutrition at birth, cold and occupation. *Human Biology*, **57**, 77–95.

Stinson, S. (1982). The interrelationship of mortality and fertility in rural Bolivia. *Human Biology*, **54**, 299–313.

Storey, R. (1985). An estimate of mortality in a pre-Columbian urban population. *American Anthropologist*, **83**, 519–35.

Stromberg, J., Peyman, H. & Dowd, J. E. (1974). Migration and health: adaptation experiences of Iranian migrants to the city of Teheran. *Social Science and Medicine*, **8**, 309–23.

Susanne, C. (1984). Biological differences between migrants and non-migrants. In *Migration and Mobility*, ed. A. J. Boyce, pp. 179–95. London: Taylor and Francis.

Syme, S. L., Hyman, M. M. & Enterline, P. E. (1964). Some social and cultural factors associated with the occurrence of coronary heart disease. *Journal of Chronic Disease*, **17**, 277–89.

Tanner, J. M. (1969). Relation of body size, intelligence test scores, and social circumstances. In *Trends and Issues in Developmental Psychology*, ed. P. H. Musser, J. Langer & M. Cogingtern, pp. 182–201. New York: Holt, Reinhart, Winston.

Tanner, J. M. & Eveleth, P. B. (1976). Urbanization and growth. In *Man in Urban Environments*, ed. G. A. Harrison & J. B. Gibson, pp. 144–66. Oxford: Oxford University Press.

Tuan, Yi-Fu (1978). The city: its distance from nature. *The Geographical Review*, **68**, 1–12.

Tyroler, H. A. & Cassel, J. (1964). Health consequences of culture change II: the effect of urbanization on coronary heart mortality in rural residents. *Journal of Chronic Disease*, **17**, 167–77.

United Nations (1975). *Status of Women and Family Planning*. Department of Economic and Social Affairs. New York: United Nations.

United Nations (1980). *Patterns of Urban and Rural Growth*. Population Studies, **68**. New York: United Nations.

Velimirovic, B. (1979). Forgotten people – the health of migrants. *Bulletin of the Pan American Health Organization*, **13**, 66–85.

Villarijos, V. M., Osborn, J. A., Payne, F. J. & Arguedes, J. A. (1971). Heights and weights of children in urban and rural Costa Rica. *Environmental Child Health*, **17**, 31–43.

Waterlow, J. C., Buzina, R.; Keller, W., Lane, J. M., Nichaman, M. Z. & Tanner, J. M. (1977). The presentation and use of height and weight data for comparing the nutritional status of groups of children under the age of 10 years. *Bulletin World Health Organization*, **55**, 489–98.

Way, A. A. (1976). Morbidity and post neonatal mortality. In *Man in the Andes*: a *Multidisciplinary Study of High Altitude Quechua*, ed. P. T. Baker & M. A. Little, pp. 147–60. Stroudsburg, Penn: Dowden, Hutchinson and Ross.

Weissman, L. L., Hood, L. E. & Wood, W. B. (1978). *Essential Concepts in Immunology*. Melno Park: Benjamin/Cummings.

Weitz, C. A. (1982). Blood pressure at rest and during exercise among Sherpas and Tibetan migrants in Nepal. *Social Science and Medicine*, **16**, 223–31.

World Bank (1981). *World Development Report 1981*. Oxford: Oxford University Press.

Yarbrough, C., Habicht, J-P., Malina, R. M., Lechtig, A. & Klein, R. E. (1975). Length and weight in rural Guatemalan Ladino children: birth to 7 years of age. *American Journal of Physical Anthropology*, **42**, 439–48.

Zarate, A. & Zarate, A. U. (1975). On the reconciliation of research findings of migrant–nonmigrant fertility differentials in urban areas. *International Migration Review*, **9**, 115–56.

Zimmet, P. (1982). Type 2 (non-insulin-dependent) diabetes – an epidemiological overview. *Diabetologia*, **22**, 399–411.

6   *In search of times past: gene flow and invasion in the generation of human diversity*

KENNETH M. WEISS

For a variety of reasons the human species does not consist of a single interbreeding population. Rather, we are an array of locally interconnected populations (demes) whose social relations with our neighbours include an exchange of mates. Often such an exchange is prescribed, but even when proscribed anthropologists have generally found that an exchange nonetheless occurs.

As mammals we are sensitized to detect morphological variation, and this has without doubt fuelled an age-old interest in explaining the nature of human population differences. Such differences show a variety of generally gradual changes over space. We know that mutation, natural and social selection, and genetic drift are important factors in generating this diversity. However, in the relatively brief period of human history, it is probable that the movement of genes from group to group, over time, has played a major role in the pattern of diversity visible today.

Genes can move from group to group via local mate exchange, or by the expansion of groups from their home ranges, if necessary replacing previous inhabitants of their new territory. Neither of these is clearly conveyed by the term 'migration'; therefore, in this paper the process of mate exchange will be referred to as 'gene flow' and the process of population expansion as 'invasion', the latter because the connotation of contested immigration is an important and relevant one. Following Cavalli-Sforza (e.g. Wijsman & Cavalli-Sforza, 1984) the term 'demic diffusion' will be used to refer to a mixture of the two, that is, expansion with incorporation of the invaded population into the resulting gene pool.

The differences between gene flow and invasion are fundamental in the theoretical models used to evaluate them; these are clearly described in Cavalli-Sforza (1983) and reviewed in some detail by Jorde (1980) and Wijsman & Cavalli-Sforza (1984). Gene flow models usually assume a more or less continuously inhabited space. The data are often represented as maps of gene frequency clines, that is, maps showing the changes in gene frequency over geographic distance. Such maps show that the further one is from a given point, the greater is the difference in gene

130

Table 6.1. *Major time divisions in human history*

Critical period (× 1000 years B P)	Major relevant events
3000–1000	Expansion of hominids from Africa into Eurasia
1000–100	Evolution of early *Homo sapiens* from *Homo erectus*
100–10	Evolution of modern *Homo sapiens*
	(and perhaps its emergence from Africa)
	(possible Neanderthal replacement)
	First settlement of Australia, the New World and the Pacific
10–3	Development of early agriculture and origin of first state societies
3–0	Mass migration/expansion of urbanized states
All time periods	Local micro-evolution

frequency. Local gene flow alone may be responsible, and is known as 'isolation by distance' to suggest that there are no actual genetic barriers, only distance, between distant populations. The best examples of human clinal maps are in Menozzi, Piazza & Cavalli-Sforza (1978), Piazza, Menozzi & Cavalli-Sforza (1981), Suarez, Crouse & O'Rourke (1985), or see Bodmer & Cavalli-Sforza (1976).

Invasion models, in this context, usually treat current gene frequency patterns in a different way. They assume that today's populations are derived by a process of bifurcational splitting from parent populations, resulting in subsequent genetic isolation of the daughter populations. In the evolutionary sense, these are 'phylogenies'. Those populations most recently descended from a common ancestor will be genetically most similar, and usually will be geographically close as well. This process can be represented as a branching tree, or 'dendrogram', although such a diagram need not be interpreted as a process of descent and isolation, and can be viewed just as a depiction of closeness of relationship. The most extensive use of dendrograms for human races is in Nei & Roychoudhury (1982).

The purpose of this chapter is to consider the relative importance of the different forms of 'migration' in the evolution of diversity at important periods of human history. These periods are outlined briefly in Table 6.1. They will be reviewed in turn.

'A la Recherche du Temps Perdu'

In his monumental series of novels, Marcel Proust argued that small cues, often of a casual or even miscellaneous nature such as a smell or a sound, lead us to recall details of the past which are inaccessible to intellectual efforts at reconstruction. We are forced to use such cues in reconstructing human population history because the record of events with which we must work is fortuitous and fragmentary. But it is not clear whether such fragments of evidence – from the fossil record, gene frequencies, or linguistics – allow us to reconstruct a truer, or less true, picture than can be reconstructed by an intellectual effort at projecting what we know from the ethnographic present about human cultures and populations onto the past. Permuting various translations of Proust's title, we are in search of 'things lost' as well as of 'times past'. We must make this search with a very incomplete set of memorabilia. Are they reliable?

Investigators attempting to infer the role of migration in the past from data collected in the present have relied on a few basic methodological strategies. These are the methods of comparative anatomy, of statistical difference computation, and the use of analogy. We can briefly summarize these before proceeding to discuss the inferences which have been drawn from their application.

Comparative anatomy

The fossil record of skeletal and artefactual remains can be used to characterize the individuals involved but, in order to relate these to other individuals or other times, comparative studies must be done showing how similarities or differences appear in the record. Similarities are usually interpreted to mean continuity and differences to mean discontinuity unless the generation of differences can be followed continuously in an area. Thus, the characteristics of material culture are used to infer cultural relationships between times and/or sites; morphological patterns of the teeth and bones, especially the skull, are similarly used but with a genetic inference added. Because of incomplete preservation and other factors, single traits are often of doubtful use. Unlike studies of comparative morphology among living peoples, palaeontology has the advantage, in principle, of a *direct* ability to assess relationships as they develop over time. However, the fossil data are very incomplete, and as a result studies of comparative culture or anatomy have taken on a special, and subjective, vocabulary. Such terms as 'total morphological pattern' or 'reminiscent' are used to justify positions which, in truth, cannot be specifically defined, and with which not all investigators agree.

These problems notwithstanding, it is unlikely that a human population will evolve morphologically in a dramatic or discontinuous way on the time-scale of only hundreds of thousands of years, so that any finding of clear discontinuity in the morphological pattern in a site over time may truly indicate the influx, from the outside, of new peoples or at least of new genes. By contrast, a rapid change in material culture may or may not imply that there has been a physical influx of a swarm of its bearers. If, however, similar styles or cultures exist at the same or earlier times in nearby areas, it seems reasonable to infer some connection.

The method of comparative anatomy has been extremely successful in the history of biology. Most taxonomies based thereon have subsequently been corroborated biochemically and genetically, and they cannot lightly be dismissed. Yet, we must be cautious about making inferences from sparse data for the very reason that it is so fragmentary. If the history of physical anthropology tells us anything, it is that one new specimen or date can overturn much of what we have believed from the totality of prior evidence, and there is no sign that this is an abating phenomenon.

Statistical difference

Numerical measures of population differentiation have been computed for many kinds of data collected on contemporary populations. These include cultural, linguistic, anthropometric and gene frequency data. For many of these kinds of data, gene flow and invasion models have been tested; that is, there have been studies applying isolation by distance and phylogenetic-tree methods. The appropriateness of the applications varies with the circumstance. However, in the majority of cases, and certainly in the most important ones, the differing methods yield qualitatively similar results; i.e. they have provided us with reconstructions of relationships which have on the whole seemed consistent with what is thought, from qualitative data, to have been the true pattern of relationships. Where problems have arisen, these have usually been (1) in the interpretation, acceptance, or rejection of statistical anomalies and outliers, and (2) in the interpretation of how the mathematical process modelled can be related to the known physico-historical facts.

Argument by analogy

A third kind of argument proceeds by analogy. This view takes observations of the behaviour of populations living today in ways which *seem* similar to what has been inferred archaeologically and assumes that the same processes applied in the past. From those assumptions, reconstructions can be made of what we believe occurred, consistent with the available palaeontological and archaeological record. For example, we typically infer the characteristics of territoriality and mate exchange among past hunter–gatherers from those observed today. This is a kind of uniformitarianism which may or may not be correct, since the present may be a deceptive representation of the past (e.g. Howell, 1976; Gould, 1982; May, 1984) and the 'lessons' we learn from primitive peoples (Neel, 1971, 1984) may be faulty ones. Nonetheless, arguments by analogy are often very powerful in evolutionary biology, and when inferences derived from metric data run counter to known primate or tribal cultural/population behaviour there is a reasonable burden on the author at least to note this, if not to offer an explanation.

In regard to migration, the argument by analogy has to do with the ability of populations, at various levels of culture, either to expand or move systematically over continental distances; to repopulate inhabited areas, even with, but especially without, extensive admixture; or to displace entire regional populations. In reconstructing the role of migration in human history it is critical to such questions concerning hunter–gatherers, Neolithic peoples and early agricultural states. Various reconstructions that have been offered have depended on extremely powerful abilities at these levels to move and replace populations. A great deal is known about inter-demic gene flow and mate exchange in contemporary human populations at all levels of culture as well as similar findings from other primate and open-country mammalian societies. Assumptions about the genetic isolation of human populations should deal explicitly with this knowledge.

Non-identifiability

Models of phylogenetic relationship are primarily derived from the study of biological taxonomy and are most reliable when applied to relationships between species or higher taxa. Such groups have been genetically completely isolated for long periods and, though their initial separation may or may not have been a discrete event, it is reasonable to investigate the rate at which differences accumulate after their isolation.

This is not the situation in regard to 'tree' treatments of human diversity. Human beings all belong to one species, and the lengths of separation of extant human populations, even under the most stringent assumptions of isolation, have been very small relative to the time thought to be required for evolutionary speciation.

The amount of genetic divergence between groups is a function of genetic drift, gene flow, mutation and natural selection, and values for all of these factors must be introduced into any model of relationships which is not a purely descriptive one (some of the parameters, such as gene flow, might be set to 0.0). Because of this, it is generally possible to fit the same data adequately to different models – to gene flow as well as invasion models – by a suitable set of parameter values.

As a result, there may typically arise the problem that the different models, and combinations thereof, are all compatible with the data. Technically, on the basis of statistical tests, it is not possible to distinguish between them, unless one can constrain parameter estimates by external sources of estimation. For example, it is often necessary to make assumptions (usually expedient ones) about mutation rates, selection, time or effective population size (e.g. Crow & Kimura, 1970; Nei, 1978; Jorde, 1980; Nei & Roychoudhury, 1982; Wijsman & Cavalli-Sforza, 1984). Under these conditions, the essential factors are those which are externally derived – but these have usually been given very short shrift. The assumptions needed are often difficult to justify.

Concluding remarks

In search of times past, one would like to know just where one was in those times. For the novelist Proust, it was Paris or Combray; he could fix his images in place and time by his certain knowledge that he had his literal origins there. For the anthropologist wishing to reconstruct human population history, knowledge is not so secure. Nonetheless, given the available methods, there are considerable data from the remote as well as recent past from which to work. While our conclusions can rarely be fixed with firmness, we may nevertheless have confidence that we can reconstruct our past with sufficient clarity that much of what we infer is correct.

Initial settlements: unchallenged invasion

If the currently available evidence is reliable, it seems clear that hominids initially evolved by a process of adaptive radiation into a series of contemporaneous species cohabiting much of Africa. It is likely that there was competition between members of the same species as well as between individuals of closely related species, but it does not seem likely that there was any large-scale population movement, and for a period of a few million years genetic microdifferentiation was probably driven by the local population dynamics among territorial ranges, with a *modus vivendi* among competing hominid species. This is what we generally observe today among African savannah primates. Stability is indicated by the long-time contemporaneity of hominid species over much of Africa. Somewhat different circumstances, however, attended the initial peopling of the rest of the world.

Eurasia

The evolved hominids of 'australopithecine' or early 'erectus' type, who first expanded out of Africa about one to three million years ago, faced little or no competition either from ecologically similar mammals or from other hominids (the only other hominids which may have been there at that time would have been gigantopithecines, who probably would have occupied very different niches). We know little of this expansion, save that hominids are found in extreme north-east Africa at about three million years ago, in south-east continental Asia by about one to one and a half million years ago, and in colder parts of Eurasia by about a half million years ago (Howells, 1980; Wolpoff, 1980).

While the population dynamics of the first occupation of Africa were probably more savannah primate-like than hunter–gatherer like, a higher cultural level seems to have accompanied the expansion into Eurasia. In any case, expansion by intrinsic growth and demic budding is almost surely the mechanism by which this occurred. There is no reason to doubt that people usually remained within a local, territorial range, and that individuals or families gradually moved to adjacent territories. There must have been strong constraints placed upon migration, such as by the location, movement and availability of food sources including animal herds. It seems unlikely that groups typically left their natal range on long one-way treks into the deep hinterland, vacating territory behind them, as the term 'migration' would evoke.

This is not to deny the exploratory urge to our ancestors. However, because of their lack of knowledge of very long distances, their social ties to their home area, and their adjacent territory either empty and suitable for settlement or occupied by others, it seems unlikely that individuals or families would truly migrate in the sense of the word today. They might move many kilometres away from their parental deme, perhaps even passing through some other local demes. But they are unlikely, as a rule, to have had any motivation for truly abandoning the area in which all their kin, potential mates, known resources, and the like were located. They would generally have kept in touch. This, at least, is my framework for interpreting the data.

Similar demic expansion led to the invasion of the colder reaches of Eurasia, though at a later stage in cultural and biological evolution – certainly without serious competition from animals. This process of unimpeded expansion stopped when the Arctic had been settled; that is, it continued until roughly 20 000–40 000 years ago.

It can be seen even from the very earliest Asian fossil specimens that there are biological affinities across the wide expanse of Eurasia into Africa, both in metric and morphological traits (Wolpoff, 1980). This is striking by the time of *Homo erectus*, when the worldwide distribution of skeletal remains shows a high degree of similarity across the entire species range (Wolpoff, 1980; Howells, 1980). Subjectively, at least, there appears to be a pattern of sufficient continuity to suggest enough gene flow across this area to maintain a single species, but enough isolation by distance to have established stable local patterns of diversity.

Though the distance between the populations at the extreme ends of Eurasia is several thousands of kilometres, even a slow trickle of gene flow, by way of demic mate exchange similar to that which occurs in savannah mammals and human hunter–gatherers, can in principle be sufficient to keep continuity over a long time period (Weiss & Maruyama, 1976). At the early *erectus* stage of human evolution and culture, it is likely that there were rather stringent limits as to the kinds of niches which could be occupied (Wolpoff, 1980). As a result, there may have been somewhat restricted migratory corridors along which most early human expansion took place. Assuming local band exogamy, which is not unreasonable, the exchange of mates would have occurred back and forth among adjacent demes on this corridor, resulting in rapid gene flow along it.

The Americas

On the basis of the best and most reliably dated archaeological material, using accelerator mass spectrometry to date minute amounts of carbon, the New World was first inhabited, via the Beringian land bridge, at some time around 15 000–20 000 years ago; the oldest reliable material may be the roughly 19 000 year date at the Meadowcroft rock shelter in Pennsylvania (USA), several 12 500 year dates in South America, the Clovis-point sites in the United States South-west dated at about 12 000 years ago, and comparably dated sites in Alaska. The earliest expansion into the New World was obviously an unimpeded incursion by anatomically modern *Homo sapiens*. These individuals appear to have been coastal resource and/or large game hunters (see Laughlin & Harper, this volume, Chapter 2). As evidence from Tierra del Fuego indicates, habitation by about 10 000–12 000 years ago (Griffin, 1979), the expansion by demic budding and intrinsic growth into the then-habitable parts of the New World must have been almost explosively rapid. This again suggests that there may have been a channelled-migration effect along rather narrow corridors and effectively speeding the north–south component of expansion. Expansion into the Amazon and Andes seems to have occurred later.

Australasia

The initial peopling of continental South-east Asia occurred without hominid opposition by around one to one and a half million years ago. The first movement into Australia required more advanced technology, presumably some form of watercraft, and appears from the fossil evidence to have occurred at least 40000 years ago (Wolpoff, 1980; Wolpoff, Wu & Thorne, 1984; Habgood, 1985). While we know there was no opposition to human settlement initially, we do not know the dynamics of the settlement pattern. It seems likely that expansion was demic and local, fuelled by intrinsic local growth at the frontiers. It now appears to this author that only one wave of immigration was necessary to generate the observed aboriginal variability (Kirk & Thorne, 1976), and the genetic (Kirk, 1979), as well as skeletal (Wolpoff et al., 1984; Habgood, 1985; Wolpoff, 1985) relationship to South-east Asians is clear. Laughlin & Harper (this volume) offer a somewhat more cautious interpretation, leaving open the question of how many immigration waves may have occurred; obviously, more data are needed.

The settling of Melanesia, especially of New Guinea, is less clear. The earliest archaeological record in the highlands is about 26000 years ago, but there are no contemporary lowland sites. Later, the coast was settled by South-east Asian and Melanesian peoples (Howells, 1973a; Bellwood, 1975). Without a better archaeological record, we must assume that these settlers met little resistance, but this cannot be argued with any degree of confidence. In any case, it appears that there are several distinct population groups on New Guinea, some more closely related to East Asians and others to Australians (Howells, 1973a; Kirk, 1979), and these peoples currently occupy different ecological zones. Hill et al. (1986) show that a haptoglobin pattern is held in common between Mongoloids, Australian Aborigines, and Melanesians from different parts of Papua New Guinea, Vanuatu and New Caledonia.

The Pacific islands were clearly settled by sea-going, small groups of colonists, and this in the last 6000 years (Howells, 1973a; Bellwood, 1975). In general, each island was occupied by one such group, immigrating from another island closer to mainland Asia. Here, colonization is clearest, though it may be more accurately thought of as a seeding, rather than invasion, of each island. Certainly it involved intentional long-distance movement, as may also have existed among the earliest Australian settlers. As a result of this pattern of settlement, there is a hierarchy of times since habitation and a systematic pattern in the biological data from the Pacific (Kirk, 1979), among other things showing that Polynesians probably arrived via a route which passed through island Melanesia (Hill *et al.*, 1985).

When present gene frequency or morphological data are used, they reflect at least indirectly what archaeological evidence tells us was the pattern by which large land areas were inhabited. This is summarized, for a sample of worldwide genetic data, in Fig. 6.1. We assume that migration

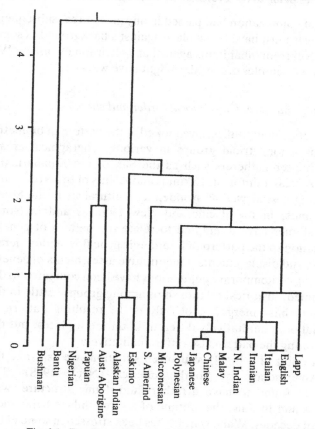

Fig. 6.1. Dendrogram of genetic distances among a selection of populations from around the world, showing their relationships. Scale shown is in arbitrary units for reference, but is proportional to genetic distance. (Redrawn with permission from Nei & Roychoudhury, 1982.)

was by intrinsic growth and demic expansion, the only kind of long-distance expansión one would expect to be regularly possible by hunter–gatherers. As genes are spread across large areas in a salient which derived from a single point of origin, deme by deme, gradients of gene frequency result. These reflect the original gene frequencies, modified by genetic drift (and, if data are assessed long after, perhaps by natural selection). Importantly, however, they may also have been seriously modified by events subsequent to the initial settlement.

Population dynamics in the middle period: gene flow in inhabited territories prior to the evolution of state agriculture

Once the front of colonization had passed in any given area, subsequent population changes must have taken place against a background of some competition with current inhabitants against any intrusion by others. We can consider some examples of how this might have worked.

Tribal microdifferentiation: local 'order and chaos'

In the past 30 years, many studies have looked at the pattern of biometrical relationships among tribal groups in various geographical areas. These include hunter–gatherers such as the San of the Kalahari, the Pygmies of the African rainforest, Eskimos and Aleuts of the Arctic, and Australian Aborigines, as well as swidden agriculturalists in the South American lowlands, in the Pacific and New Guinea, and in North America. The studies have attempted to relate the pattern of genetic microdifferentiation to the pattern of interdemic gene flow and/or demic fissions, fusions, and displacements. Quantitative assessments of genetic differences among contemporary populations have been compared to the known, or assumed, historical relationships among groups. Little in the way of generality has emerged. Not only is the resulting pattern of microdifferentiation quantitatively different in different places, but the processes generating the pattern also differ qualitatively.

As a case in point of a highly structured situation, we can take the Yanomama Amerindians of South America (Neel, 1978). The Yanomama were found to have a rather fluid demic structure, with frequent fissions and fusions, the lifetime of a given village being measured in terms of decades. While warfare has been stressed in some of the literature on this tribe (e.g. Chagnon, 1968), other phenomena can also lead to a loss of village identity; these include reaching a critical size, resource stress or exhaustion, family squabbles, or simply casual changes in membership. In any case, fluid demic structure is probably common among horticultural populations (Smouse, 1983; Neel, 1984).

Genetic relationships among Yanomama villages generally reflect their historical affinities over the very recent past (Smouse, 1983; Neel, 1984). Group fissioning and fusioning usually take place within local areas occupied by the tribe, and the roughly random occurrence of these events over time leads to clusters of related villages in those regions. This kind of local fissioning occurs sequentially over time; thus, the villages produced by one fission themselves later fission independently. This produces a 'tree' of fission products, that is, of local village relationships. The chance element in this process increases some gene frequencies at the expense of others. Since this is random within any given region, the pattern of multiregional or multitribal gene frequencies shows no obvious clines or smooth trends over space. Hence, there can be local clustering of related villages but an absence of such patterns on a broader scale.

To use an apt characterization by Ward (1976), what occurs is a pattern of 'order and chaos'. The 'chaos' is the unpredictable fission/fusion process, but this produces 'order' in the relationships among local demes. Genetic or anthropometric (Spielman & Smouse, 1976) reconstruction of relationships can be successful on the local scale, and will reflect recent history. The more distant past is clouded in the 'chaos' of random genetic 'noise'. In the Yanomama, as in many other tribes, this general correspondence of local relationships can also be found in linguistic, dermatoglyphic, geographic and anthropometric data (Spuhler, 1972, 1979; Spielman, Migliazza & Neel, 1974; Chakraborty *et al.*, 1976; Jorde, 1980; Smouse, 1983).

The highly structured local microdifferentiation of the Yanomama results from their having quasi-permanent villages which come and go as units. This may be due to the relationships between the labour needs of swidden agriculture as a factor concentrating population, and the limited productivity of the land leading to fissions when population gets too large (Smouse, 1983). This can even occur among hunter–gatherers (e.g. Szathmary, Ferrell & Gershowitz, 1983). But not all tribal peoples are like that. In many areas of the world, especially among hunter–gatherers, there is a much more fluid demic structure, where *individuals* come and go frequently. The Pygmies (Cavalli-Sforza, 1971) and Bushmen (Harpending & Jenkins, 1974) in Africa, and Arctic Eskimos (Murdock, 1934), may be examples. In such groups there is very little genetic differentiation within regional areas. However, different regions can themselves be considered undifferentiated 'clusters' relative to other regions; that is, trees of relationship can be constructed among regions, but not among the temporary demes within them. Here, as with the more structured case, there need not be smooth clines of gene frequency over large geographic areas.

In tribal cultures of the kind under discussion, local demes have about 50–200 people, all of whom work the local land or hunting territory for their subsistence. For them, there is nothing which can correspond to an army of warriors who can be spared for distant conquest. Such peoples do not own land occupied by others or administer if from afar. Hence, it is unlikely that the members of one deme, and their descendants, will *systematically* expand over a large area and displace its previous inhabitants in the political sense usually evoked by the term 'invasion'. However, by virtue of a particularly bellicose and capable headman, or some advance in competitive technology, at any given time there may be local demes which are expanding to dominate a larger area at the expense of other villages. Eventually, one would expect either a response in kind from the victimized villages, loss of momentum and/or borrowing of the technology (e.g. by imitation or mate exchange). Some villages or even whole tribes may virtually disappear, be displaced long distances, or be fragmented into isolated remnant pockets of population. Foci of expansion would continually ebb and flow across a geographic area in more or less stochastic fashion. No one focus would dominate except temporarily.

Genetically, however, such random genetic hot-spots would, over time, lead by chance to the increase in frequency of a subset of the original alleles at the expense of others. If we could trace the history, we might see that in fact the genes originally found in one of the centres of local expansion would end up in much higher frequency over a large area. This is the larger scale effect, over time, of the local order and chaos. Smooth gene frequency clines, however, need not result.

In sum, what occurs at the tribal level is highly dependent on its geographic and cultural context, but seems generally to lead to local genetic 'patchiness' – irregularities in gene frequency patterns. Rough correspondence between gene frequency and geography, on a continental scale, usually still occurs because the local regions do share ultimate common ancestry; but such tendencies may only be very general ones. It is important to note that maps of gene frequency clines, which involve a considerable amount of interpolation (in areas for which there are no data), may give a misleading appearance of smoothness over very large areas, masking true patchiness which may exist. On the other hand, it must be remembered that broad and/or smooth continental trends may exist, and may reflect selection (e.g. the effects of climate on north–south genetic clines (Guglielmino-Matessi, Gluckman & Cavalli-Sforza, 1979;

Piazza *et al.*, 1981; O'Rourke, Suarez & Crouse, 1985). They may also reflect genetic barriers such as mountain regions and deserts, or they may reflect the original settlement history of a recently settled area such as the Americas. Migration is not the only smoothing process. From gene-frequency data alone, the non-identifiability problem prevents us from determining what combination of these factors actually applies.

Advanced pre-state agricultural populations: demic diffusion over large areas

The pattern of interaction at a more advanced pre-state agricultural stage of human culture probably differed greatly from place to place, but in terms of the remembrance of such past things from today's data, we may again take a case in point, namely that which has been studied in the most detail. This is the spread of neolithic culture into parts of Europe.

For over a decade, Cavalli-Sforza and others have related the pattern of genetic variation in Europe to the spread of Neolithic farming (summarized in Menozzi *et al.*, 1978; Sokal & Menozzi, 1982; Ammerman & Cavalli-Sforza, 1985). Farming appears to have been introduced into Europe from the Near East. They argue from the genetic as well as archaeological evidence that there was a physical expansion of Neolithic *farmers* as well as farming into Europe and that these individuals, while admixing with the local hunter–gatherer population, bore a sizeable fraction of the genes ancestral to today's Europeans; in other words, they believe that this was an instance of large-scale demic diffusion.

Their argument is an important one because it documents the ability of relatively simple cultures systematically to occupy a previously inhabited territory by the displacement of the aboriginals, on a continental scale, and to do so with only a modest technological advance compared to confrontations which have occurred subsequently. Such a replacement would be the best documented case of pre-state cultures expanding into inhabited territory on a large scale. By comparison with the horticultural peoples of Amazonia, whose population density and technology are not that superior to hunter–gatherers, in Europe there seems to have been a much greater technological advantage and higher population density among the farmers. Because of its importance, therefore, the European story will be reviewed.

Two kinds of data have been used to evaluate the nature of the population processes responsible for the spread of agriculture in Europe. First, using multiple genetic marker loci, Cavalli-Sforza and several colleagues have demonstrated a major north-west to south-east multilocus gene frequency gradient (Menozzi *et al.*, 1978; Piazza *et al.*, 1981; Sokal & Menozzi, 1982; Ammerman & Cavalli-Sforza, 1985). This gradient supports the wave of advance, or demic diffusion model. Corroborating gene frequency patterns in the HLA system have also been found in Europe (Sokal & Menozzi, 1982).

It is impossible from a gradient of gene frequencies alone to assign a true historical explanation, since other causes, such as selection, could also produce the pattern; this again is the non-identifiability problem. However, in the case of Europe, other evidence makes the case a more convincing one. This is the archaeological evidence for an incursion of Neolithic farmers into territory occupied by Mesolithic hunters, along an area from the Near East, through Eastern and Central Europe to Holland in the period from 9000 to 5000 years ago. The prior Mesolithic inhabitants had apparently been limited in the types of ecological zones in which they lived, and had left river valley areas of loess soil relatively vacant – at least, if they, or their game, used such areas, they were displaceable. The Neolithic peoples moved into these areas with a front of advance which moved at a rate of about 1 km per year. As reflected in similar house-structure and pottery (the 'Bandkeramik') styles, this was rapid enough to prevent marked cultural differentiation over this entire distance. It must be noted, however, that not all archaeologists agree with this interpretation. Among other things, there is disagreement that the Bandkeramik can be interpreted as a single advancing, coherent cultural style (Chapman, 1985).

Because there is a gene frequency cline which fits this expansion pattern, there must have been genetic differences between the Mesolithic and Near Eastern peoples before the expansion. More importantly, for a cline to be established under such circumstances, there must be some admixture between the invaders and the inhabitants (e.g. Menozzi *et al.*, 1978; Ammerman & Cavalli-Sforza, 1985). At each point in the advancing wave, incorporation of 'Mesolithic' genes occurred in the farming population. By the time this wave reached north-west Europe, it seems likely that only a fraction of the original Near Eastern genes remained, the rest of the gene pool being the result of the increased level of indigenous genes incorporated into the advancing front. Even in this kind of demic diffusion, therefore, when the indigenes did not borrow the new culture *en bloc*, by the time the front reached its limit it was really a Near Eastern culture implemented by largely European genotypes.

This kind of displacement may explain the presence of isolated relict populations in many parts of the world, for example the San (Bushmen) and Pygmies in Africa, the Basques in Europe, the Ainu in Japan.

Population movement on a comparable scale probably had occurred in Aztec and Inca areas of the Americas also.

The New World: demic diffusion of uncertain origin

North America has been settled for some 15000–20000 years. It is interesting to see what may be said of gene flow versus invasion based on comparative population data.

The pattern of initial settlement of the New World has not been worked out to everyone's satisfaction as yet. There appears to be a general (though incomplete) consensus that three somewhat distinct groups of American aboriginals can be distinguished, both at present and in the archaeological record. These are the descendants of the Palaeo-Indians, and Athapascans, and the Eskimo–Aleut (Laughlin & Harper, this volume; Laughlin, Jorgensen & Frohlich, 1979; Harper, 1980; Turner, 1983). Some studies, based on skeletal and genetic data, suggest that these people are more like each other than they are like Asians (e.g. Szathmary & Ossenberg, 1978; Szathmary, 1979b, 1981, 1984), possibly representing peoples who differentiated *within* the New World. This view is advanced in the current volume by Laughlin & Harper. Other investigators, using genetic as well as dental data, infer that these three separate peoples represent separate physical immigrations from Asia (Harper, 1980; Sukernik & Osipova, 1982; Turner, 1983; Williams et al., 1985). Biometrical data support the cultural and linguistic data only in a general way, mildly suggesting a three-part distinction among the American peoples. The consensus view has had the Palaeo-Indians coming across Beringia in pursuit of large land mammals, and the Athapascans and Eskimo–Aleut arriving subsequently. Fig. 6.2 is a dendrogram for a selection of American and Asian populations to show the general nature of the relationships among the various subpopulations; the lower major branch contains only 'Palaeo-Indian' descendants.

The amoeba of consensus can move in unpredictable directions, and in this case two kinds of recently developed data indicate that there may be complications. There are several consistent dates of about 12500 years ago in South America (e.g. Lynch et al., 1985) indicating arrival of people earlier than in North America. Also, the rockshelter at Meadowcroft, Pennsylvania, may provide the best evidence for Palaeo-Indian habitation at about 20000 years ago (Adovasio et al., 1983). If this latter date is correct, then there would have been a period of a few thousand years during which peoples in what is today the United States were effectively cut off from their northern kin by transcontinental glaciers in Canada

146   K. M. Weiss

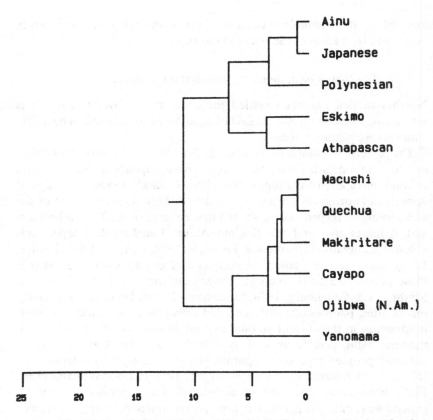

Fig. 6.2. Dendrogram showing relative genetic distances among selected native American and Asian populations. The separate groups of Amerindians and their relationship to an ultimate Asian origin can be seen. Scale shown is in arbitrary units for reference, but is proportional to genetic distance. (Redrawn with permission from Nei & Roychoudhury, 1982.)

(Fladmark, 1983; Szathmary, 1984). During that time, the differentiation of the Eskimo–Aleut and the Athapascans may have occurred in the isolated habitable part of the Arctic New World (e.g. Szathmary & Ossenberg, 1978; Szathmary, 1979a, 1981, 1984), although Laughlin & Harper (this volume) raise the question of where the Athapascans were during the thousands of years since Eskimo–Aleut separation if this occurred in the New World.

There is reason to suspect that the notion of a steppe-like Beringian land bridge may be incorrect and that instead the terrain was relatively barren and uninhabitable (Colinvaux & West, 1984; Colinvaux, 1985;

Laughlin & Harper, this volume). If so, then immigration into the New World may have been constrained to a channel of flow across the southern Beringian coast. Differentiation of American peoples could have occurred entirely in the New World rather than in separate waves of immigration. For example, the descendants of the first coastal immigrants could have led to one penetration, along river systems, to become the Palaeo-Indians, while the remainder developed into the coastal specialized Eskimo–Aleut and, in the tundra areas of Alaska, the Athapascans. Of course it would also be possible for there to have been a series of pulses of peoples, or of channelled gene flow, along the coast over a period of time, with genetic differentiation occurring in Asia. The genetic data are insufficient to resolve this at present. It is difficult to support a rigid typology among American native peoples in various respects, or to accept that such divisions would have remained completely separated in the thousands of years since they have clearly shared the northern part of North America. Many aspects of American ethnohistory remain to be resolved.

The genetic data show the north–south migration track of initial habitation of the New World (Suarez *et al.*, 1985), although this has also been interpreted as reflecting climate (O'Rourke *et al.*, 1985). Many unresolved problems remain. If all differentiation has occurred subsequent to the colonization of North America, then the Athapascan and the Eskimo–Aleut peoples may have differentiated in a kind of local order-and-chaos environment, where some geographic or cultural isolation and niche divergence took place, but where demic diffusion or displacements did not occur. On the other hand, if there have been multiple independent settlement waves across Beringia, especially if all have been post-15 000 years – suggesting that there was no period of glacial isolation of the Arctic from the south (e.g. Laughlin *et al.*, 1979; Griffin, 1979; Turner, 1983; Owen, 1984) – then one must assume that the second and third waves were demic diffusions or invasions into territory inhabited by the Palaeo-Indians. At the time, all these peoples would have been hunter–gatherers of various sorts. The Eskimos may have had an advantage due to their extreme niche specialization (sea-mammal hunting). For the Athapascan ancestors, we do not know on what advantage their invasion would have relied.

Regardless of how the initial peopling occurred, it seems clear that in very recent pre-Columbian times, there was an expansion of Athapascans into the United States South-west. They are today represented by the Navajo and Apache. This may have occurred across the sparsely occupied great plains, involving rather little opposition, but the archaeological evidence gives no support for that. Alternatively, the Athapascan salient, today represented by their presence along the north-west coast, may have followed a totally western path; there is some evidence for this in the existence of California and British Columbia Athapascans. This path, one might surmise, would have involved expansion at the expense of indigenes; perhaps the Athapascans had the means of a higher population density, but the archaeological record does not show this and it is somewhat difficult to understand the mechanism of their expansion. In that connection, it is clear that between the north-west part of Canada and Arizona, most Athapascans have subsequently been replaced, displaced, or absorbed by Palaeo-Indian descendants. However these events took place, it is not on genetic but on cultural and archaeological evidence that we must rely to resolve the true historical events.

Recent human history: invasion and admixture by state societies

The large-scale population movements of the past few thousand years of human history are well known and of a dramatic nature. Agricultural states have large territorial populations but ones which contain large numbers of specialists who are *not* tied directly to the land. Unlike populations at earlier cultural levels, such populations can (1) mount and supply large armies of individuals who can travel long distances in expeditions of territorial conquest, (2) spare large numbers of individuals as colonists who may seed the conquered territories, and (3) maintain dominance over distant lands by means of hierarchical administrative systems and long-distance communications.

These attributes permit a shift from expansion by demic diffusion to what we would easily recognize as true invasion. Such invasion can lead to the rapid effective replacement of the genes of aboriginals by those of the invaders. This occurs by (1) the killing of the inhabitants in military conflict, (2) the killing of the inhabitants by new pathogens to which they are vulnerable, (3) the movement of numbers of colonists large in proportion to the inhabitants, and (4) the higher intrinsic growth and settlement density made possible by the importation of advanced cultures. All of these are new to the development of state societies, which are able rapidly to replace one population with another even on a continental scale without relying on large amounts of admixture with the original inhabitants.

To date, this long-distance colonization has led to great discontinuities on the contemporary genetic map. For example, the Americas now have, juxtaposed and intermingled, genes of the most disparate peoples

(Europeans, Africans, Amerindians). In the long run, however, if this continues we can expect a worldwide homogenization of the human gene pool. We will now briefly review the history of human migration at the level of contact between pre-state and state populations, between advanced and lesser states, or among advanced states.

Europe

In addition to invasion from external populations such as the Mongols and Moors, the recent history of Europe comprises many large-scale internal population movements. The expansion of the Roman empire, its retreat in competition with peoples of northern Europe, and the many invasions of the British Isles are examples of movements by substantial numbers of people who colonized areas into which they moved. While local genes remained well represented, these movements differed from demic diffusion in that they were not simply the result of intrinsic growth at an expanding frontier, but specific expeditions of distant conquest. Another expansion which was neither demic diffusion nor military was that of the Jews into Europe. It represented a kind of expansion which has become possible only because large numbers of people are not tied to their land; indeed, the Jews served many economic functions which encouraged Europeans to provide a place for them (interestingly, their descendants have reinvaded their original homeland). The European genetic map reflects its history of population movement, and it is curious that the much earlier Neolithic expansion should still be visible after such turmoil.

Africa

The agricultural Bantu spread dramatically across Africa in the not too distant past, about 2000–3000 years ago (e.g. Cavalli-Sforza & Bodmer, 1971), displacing the non-Bantu populations who had been there for many thousands of years. Bantu Africans eventually came to develop advanced social organization, including states, but when they were confronted with later European colonists the results were dramatic. However, the majority of the African continent remains genetically African, even if European-drawn territorial boundaries and the effects of outside colonization are altering the pattern of intertribal gene flow. At the same time, Africans were, in a way, becoming a major invasion force in the New World, and their genes now are a major segment of the American gene pool.

India

Many parts of India seem to retain indigenous tribal peoples with long history in their current location; the distant ancestral relationships of these peoples are unclear (Roychoudhury, 1984). However, there is clear evidence, reflected genetically and historically, of populations flowing into the north-west and east of India. Much of this has occurred in the era of the agricultural state, expansions occurring within central India by indigenous agricultural peoples, and from outside India from the west. Indians in the east of the country are clearly closely related to Mongoloids, whereas those in the bulk of the South, West, and North show their 'Caucasian' ancestry genetically as well (Roychoudhury & Nei, 1985). Perhaps because it has been such a crossroads, many geneticists and anthropologists do not think Indians should be classified as a 'major' racial group, but rather as a series of 'indigenous' racial stocks plus a variable admixture of Caucasians and Mongoloids. The numerical bulk of the population represents the differential growth of the aboriginal populations who have shared ancestry with Caucasians and Mongoloids. However, we know that even after this expansion the tribal peoples are still represented by at least 450 groups, which represent about 7% of the population (Roychoudhury, 1984).

Asia

Today, much of eastern Asia is inhabited by Chinese, Japanese, and related peoples. This group of people differs genetically and morphologically from many Asian 'aboriginals' – for example, tribal populations of South-east Asia, the New Guinea and Philippine natives, the Taiwan aboriginals, the Ainu, and others. These aboriginal peoples seem to have been displaced by the expansion of the 'Chinese' peoples from a source in the Great Bend area of the Yellow River.

This is supported by archaeological evidence, which shows that a north-east expansion of agriculture-bearing peoples from South-east Asia has occurred in the past 10 000 years. It is important to remember that 10 000 years ago the mainland of Asia was populated by tribal populations without, so far as we can tell, a domination by what today are Chinese people. However, there are reasonable gene frequency (Kirk, 1979; Piazza et al., 1981), skeletal (Wolpoff et al., 1984; Wolpoff, 1985), and dental (Brace, Saho & Zhang, 1984; Turner, 1983) gradients relating today's peoples from Australia to Siberia.

Australia

Australia has been nearly fully repopulated by Europeans, the remaining aboriginal genes generally becoming a small fraction of the gene pool there. Yet despite the cataclysm attending European colonization, there has been some admixture, and there are still aboriginal genes on the continent.

The Americas

The invasion of North America (United States and Canada) by Europeans and by Africans has greatly rarefied Amerindian genes in the gene pool. Certainly, as in Australia, this must count as a true invasion.

The story in Latin America was different. While many Latin American populations are predominantly European or European/African in nature, much of Central and South America is today composed of the admixed descendants of the European and American societies which came into contact when the Spanish invaded the New World. Unlike in North America, the density of populations in places like the Valley of Mexico may have been such that the Spanish intended initially to import men, who took Amerindian wives, rather than to import whole families, as done by the English and French to the north (many of whom were escaping persecution, not seeking gold). For whatever reason, intermixture on a large scale began early, and today American genes are still a major fraction of the Latin American gene pool, and to a considerable extent they are randomly spread throughout the predominantly European population. In recent years, there has been a clear spread of these peoples northward in North America as a result of very large-scale long-distance migrations.

Discussion

It is clear that the confrontation between advanced, large agricultural-state populations leads to a diffusion of their genes on a large scale across great distances and this expansion need not be into areas contiguous with the state itself. This may be just the large-population extreme of the process we have been reviewing. It may be reasonable to generalize by saying that when two cultures confront each other, the rate at which one can expand at the expense of the other, genetically and geographically, is a function of the degree of difference between the two technologies and their consequent population densities. At hunter–gatherer levels, populations are small, processes slow and geographically irregular, and invasion less effective than gene flow in determining genetic differentiation. As culture and population densities advance, movements are greater, and so is the relative importance of invasion. In addition, for social reasons in complex societies, some subpopulations systematically may grow and expand more than others, changing the genetic map without any interpopulational movement. This may be occurring, for example, in the current growth of the Black and Mexican–American segments of the United States. Essentially these are all rate phenomena which are functions of the kinds of culture involved.

Some unanswered questions remain. How much of the current pattern of biological diversity is the result of the long-term action of localized processes in a species widely dispersed in space, perhaps having accumulated over 500 000 or a million years since *Homo erectus* emerged from Africa? How much may reflect patterns caused by demic diffusions during the development of more sophisticated extractive techniques, including early agriculture? And how much may be the overlay of the greatly accelerated processes of true displacement which have occurred just in the last 10 000 or fewer years?

These questions are difficult to answer, because adequate records have not been left in all cases of the identity of the people who inhabited areas before the occurrence of displacements which they suffered. It is to some of these issues that we now turn.

From the present looking back: the history of major human races

Genetic data have been used by several authors to reconstruct the origin of the major subpopulations of humankind, both in time and in place. This is a subject of great potential interest, but is an area fraught with problems.

To do this, the techniques of dendrogram construction have been applied to the study of population samples taken to represent the 'major' races of the human species, that is, the numerically dominant peoples of the present day (Cavalli-Sforza, 1969; Cavalli-Sforza & Bodmer, 1971; Latter, 1973; Nei & Roychoudhury, 1974, 1982; Piazza et al., 1975; Nei, 1978, 1982; Brown, 1980, as interpreted by Nei, 1982; Piazza & Cavalli-Sforza, 1983). The data used include blood groups, serum proteins, and genomic and mitochondrial DNAs. Skeletal data have also been used (Howells, 1973b; Guglielmino-Matessi et al., 1979).

The assumption that the 'major' human races are Negroid, Caucasoid, and Mongoloid (these not precisely defined) has led to the usual practice of selecting one or a few sets of gene frequency data to represent these races: for example, from Bantu West Africa and the United States Black population, from Japan, and from northern Europe. American Indians, non-Bantu Africans, and other populations are also often included. Trees have been constructed from these data by a variety of methods. The topology of the trees varies among the studies, particularly in regard to whether Caucasoids are more recently separated from Negroids or from Mongoloids (Nei & Roychoudhury, 1982; Piazza & Cavalli-Sforza, 1983; Johnson *et al.*, 1983; Nei, 1985). The kinds of differences are shown schematically in Fig. 6.3. In some instances, particularly those involving mitochondrial DNA, it has been difficult to arrange a tree which adequately suits our notions about the major races, for example

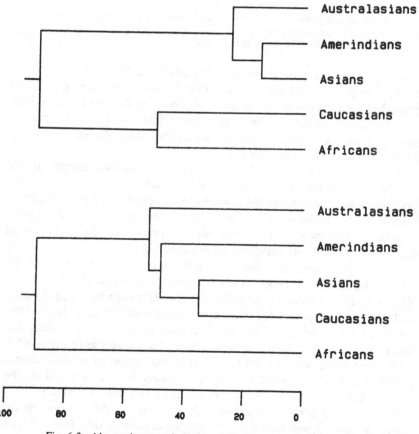

Fig. 6.3. Alternative morphologies of dendrograms of relationship among the major races of contemporary mankind. Dendrograms resembling these have been time calibrated and interpreted as phylogenetic trees. Schematic and not drawn to scale. (Based on Nei & Roychoudhury, 1982, Piazza *et al.*, 1975.)

Table 6.2. *Restriction enzyme polymorphisms at the*
β-globin region in selected human populations

Haplotype	Populations			
	Europeans	Indians	Asians	Africans
+ − − − −	107	58	189	3
− + − + +	44	28	35	6
− + + − +	15	15	5	1
− + − − +	0	0	8	12
− − − − +	0	0	2	37
Others	3	10	21	2
Total sample	169	111	260	61

Haplotype pattern is presence (+) or absence (−) of a restriction enzyme
site on the chromosome.
(Modified after Wainscoat *et al.*, 1986).

showing a chaotic pattern or showing evidence for an eastern Asian
human origin (e.g. Denaro *et al.*, 1981; Cann & Wilson, 1983); this
probably has to do with problems associated with the small and preliminary nature of the data set because as more data accumulate the pattern
seems to become more consistent with other data (Cann, 1982; Nei,
1982, 1985; Johnson *et al.*, 1983).

A recent article on this point might serve as a valuable example. In this
article, Wainscoat *et al.* (1986) have compared restriction-enzyme
polymorphism patterns at the β-globin gene region. For five restriction
enzymes, the frequencies of several haplotypes were determined for a
number of European, Asian, and West African individuals. The main
points of the results are summarized in Table 6.2.

The authors noted the distinct haplotypes which predominated the
African sample, and the generally common pattern found in Eurasians.
Their conclusion was that the three common Eurasian haplotypes, along
with the most common African haplotype, pre-dated the others as the
original human morphs. That the Eurasian frequency, or genetic distance, patterns were similar across that vast land area was interpreted to
mean that there was a recent and rapid expansion of a small source
population from an African emigrant source. In that expansion, the
predominant African (and hence an aboriginal) haplotype was lost.

If true, these arguments are consistent with the African source for
human evolution, and with a recent expansion from Africa. However, an
accompanying editorial (Jones & Rouhani, 1986) showed that if humans
emerged from Africa at around 50 000 years ago (age of first anatomically
modern *Homo sapiens*), a rather severe bottleneck in population size

Table 6.3. *Estimated separation times (× 1000 years BP) between major human races*

Caucasoid–Negroid	Caucasoid–Mongoloid	Mongoloid–Negroid	Reference
20–85	15–60	25–100	Cavalli-Sforza, 1969
<30–50	30–50	30–50	Piazza *et al.*, 1975
<35–40	35–40	35–40	Guglielmino-Matessi *et al.*, 1979
104	104	104	Johnson *et al.*, 1983
<180–360	<180–360	<180–360	Brown, 1980
50–100	50–100	50–100	Latter, 1973
53–106	20–40	45–90	Nei, 1982
113	41	116	Nei & Roychoudhury, 1982
35–75	13	35–75	Nei, 1985

These estimates come from different data sets using different kinds of data. Estimates by papers co-authored by Cavalli-Sforza and those by Nei are not all based on independent criteria. Values are averages or rounded figures.

would be required to produce the observed genetic pattern. This is not consistent with data from many other genetic marker loci.

An alternative explanation may show the difficulties involved with these kinds of reconstruction. The African data are all from Bantu individuals. The Bantu represent a rapid and recent expansion within Africa only in the past few thousand years. It might be that the Bantu source population, through mechanisms of mutation and recombination including drift or selection in the time before their expansion, developed a haplotype distribution not representative of the rest of Africa at that time. If so, inferences about the degree of difference of contemporary (Bantu) Africans from other populations, or about divergence times, may be difficult to make with any level of confidence. Indeed, a rapid divergence from a small source population of hunter–gatherers might actually lead to a much more heterogeneous pattern than is observed today.

Some authors have assumed that the dendrograms represent the true (phylogenetic) history of the populations, and have attempted to estimate the time since any two populations have been separated, that is, since they become genetically isolated from each other. The timing of such branchings can be based on many different criteria, including known historical events such as the separation of hominids from pongids, of the Amerindians from Mongoloids, or on estimated rates of molecular evolution in nuclear or mitochondrial DNA. A variety of such criteria has been used by the authors who have represented the evolution of the major races phylogenetically. Their separation time estimates are given in Table 6.3. These estimates, which themselves have large standard errors, vary from about 13 000 years to about 360 000 years or more. Nonetheless, some investigators feel that estimates in the range of 40 000 to 100 000 years have consistently resulted from quite diverse data and methods, indicating their reliability (Johnson et al., 1983).

There are fundamental problems with the use of these kinds of data to construct time-calibrated trees of human racial ancestral history. The evolving races may not have been isolated from gene flow from each other or from other populations during this period, and therefore it is unlikely that the separation time estimates can be interpreted literally in terms of the true history of the racial groups (Weiss & Maruyama, 1976).

These problems thus do not lie with the methods used but with the use of the methods, that is, with the assumptions made. The same assumptions also imply much about invasion and gene flow. While the data can be made to conform to the models, the assumptions seem to be difficult to reconcile with what is known archaeologically about human history and ethnologically about human cultures. At the very least, some explanation is called for.

Isolation of major racial groups

Within any continental geographic area, it is clear that gene flow has had a major effect on the present-day gene frequency picture (e.g. Nei & Roychoudhury, 1982). Although they had to do a considerable amount of interpolation (smoothing) to account for large areas for which there were no data, Piazza et al. (1981) have shown evidence for an east–west clinal gene frequency pattern across Eurasia, presumably also due to gene flow (isolation by distance). Nonetheless, when trees have been constructed for 'major' races, intercontinental gene flow has been assumed to be unimportant relative to the degree of isolation of the sampled populations. Representing major races only by geographically peripheral populations, as the same authors do when constructing major-race trees, essentially ignores any such clines and the genetic continuity they imply.

It is appropriate to note here that both clinal and phylogenetic treatments of the data make assumptions which must be taken into account when judging their results. Genetic data are multidimensional and cannot be portrayed on a flat paper surface without some simplifying reduction of dimensionality. Both dendrograms and clinal maps digest the data and project it onto paper using various statistical methods. Essentially, only some of the available information is portrayed. Trees treat gene frequency patterns as *discontinuous* branches, and timing the assumes bifurcation with isolation. Clinal maps, on the other hand, assume that gene frequencies change *continuously* over space, and as a result they make the assumption that missing data would have intermediate values relative to the closest sampled populations. This results in what might be a misleadingly smooth picture. Either method thus may mask an important part of the variation. There is no right method.

Separation time assignments

Even if there had been *approximate* isolation of ancestors of today's major races, a small amount of gene flow can greatly reduce the estimates of the time that isolation has existed (Weiss & Maruyama, 1976), making it seem as if a common ancestor lived much more recently. On the other hand, if the separated populations receive gene flow from other, outside, populations, their apparent separation time may be increased. If we cannot assess the relative importance of these counteracting effects, our ability to relate such separation times to the archaeological record is seriously hampered. Because of the effect of migration, Nei prefers to call his trees 'dendrograms' rather than 'phylogenies', and says that separation time estimates should only be considered as 'effective separation times' (e.g. Nei, 1978; Nei & Roychoudhury, 1982), a point also acknowledged by Cavalli-Sforza (1983; Piazza & Cavalli-Sforza, 1983). The effective time is the time it would take populations who have descended from a common ancestor to develop their observed degree of genetic diversity in isolation. The effective time is in historical error to the degree that these assumptions are not met. Even when this is acknowledged, however, authors drawing race trees have calibrated and/or analysed their trees in terms of real historical and archaeological events, giving them, after all, a literal interpretation.

Other technical problems

There are other technical problems with the use of phylogenetic trees in this context. For example, under the assumptions of bifurcation and isolation, there should be a kind of symmetry in the genetic variation among populations purported to have descended from a common ancestor. If two populations are derived from the same common ancestor, they should differ genetically from that ancestor by the same amount. However, this is not consistently observed, casting doubt on the phylogenetic assumptions (Piazza & Cavalli-Sforza, 1983). It is also necessary to guess at values for mutation rates and the population sizes of racial groups during their history (e.g. Nei & Roychoudhury, 1974; Weiss & Maruyama, 1976; Nei, 1978; Weiss, 1984). Furthermore, history might be 'erased' by selection and other non-systematic events (Livingstone, 1980), and indeed there is recent substantial evidence based on multivariate genetic data that climate may be one such selective force (Guglielmino-Matessi *et al.*, 1979; Piazza *et al.*, 1981; O'Rourke *et al.*, 1985).

Migrational implications of race-separation times

Any race separation time has migrational implications. It may not be easy to resolve which of the competing sets of implications is likely to be the most correct. However, it is reasonable that these be made explicit, rather than being latent, or unstated, assumptions of a given model. If we look at the various possible racial separation times, we can examine what the migrational difficulties of each might be.

We might begin by supposing that there has been *no* separation of the major races; that is, that over 500 000 to a million years in the Old World there was always sufficient gene flow that *Homo erectus* evolved gradually into *Homo sapiens* over this entire area. This view is, by and large, the one held by the present author (Weiss, 1984; Weiss & Maruyama, 1976). It implies that nowhere in this long time period did major rapid population expansions/replacements, without substantial admixture, take place.

Racial separation times less than this imply in one way or another a large and systematic amount of invasion. Wherever the core area was, the antecedents of today's races would have had to expand throughout the Old World to replace the inhabitants who had been there for over a half-million years. It is difficult to know what their advantage would have been, or how this occurred without there being considerable admixture. We cannot rule out some technological advantage, or even a biological one (e.g. an immunological advantage, or a chromosomal incompatibility with indigenous people), but there is no evidence even to suggest it. I say 'replace' rather than 'displace' because racial separation times are computed under the assumption that the branches evolve without external admixture. Otherwise, the time estimates are in error to an indeterminate extent as already noted.

This problem seems to me to pertain especially to separation time estimates of 40 000 years ago or less. One basis of such estimates has been the argument that anatomically modern *Homo sapiens* first appeared in Africa 40 000 years ago (Rightmire, 1981, 1984; Stringer, Hublin & Vandermeersch, 1984; Brauer, 1984), though even Brauer argues for hybridization rather than worldwide replacement). Another simply rests on the appearance of anatomically modern *Homo sapiens* at about that time across the Old World, perhaps replacing Neanderthals in Western Europe (Guglielmino-Matessi *et al.*, 1979; Piazza *et al.*, 1981, 1985; Johnson *et al.*, 1983; Stringer *et al.*, 1984; Trinkaus, 1984). But how would a band of hunter–gatherers from Africa (or anywhere else) have been able to emerge to become the progenitors of all humankind? They would have had to expand across thousands of kilometres, adapting as they went to climates ranging from tropical to arctic, and replacing peoples already themselves adapted to these areas and who surely were territorial and experienced with projectile weapons. Australia was apparently already settled by 40 000 years ago by *Homo sapiens* who bear skeletal affinities to past as well as present Asians (Wolpoff, *et al.*, 1984; Habgood, 1985; Wolpoff, 1985). Further, the expanding wave would have had to reach the New World by about 15 000 to 20 000 years ago. This all requires a very rapid expansion and presents acute problems for arguments that the races are only 20 000 to 30 000 years old.

It is important to recognize that there appears to be morphological continuity in the features of the skeleton (primarily the skull and dentition) in many parts of the world from *Homo erectus* to *Homo sapiens*; that is, there is a 'racial' continuity in (1) fossil eastern Asians, including Australians and Indonesians, and today's eastern Asians (Kirk & Thorne, 1976; Wolpoff, 1980; Wolpoff *et al.*, 1984; Habgood, 1985; Wolpoff, 1985), (2) western Asians where there is evidence for much *in situ* evolution (Trinkaus, 1984), (3) distinctive African characteristics (Rightmire, 1981, 1984; Brauer, 1984) with no evidence of major population replacements perhaps even for a hundred thousand years (Rightmire, 1984), and (4) in central Europe from Neanderthal to modern humans (Smith, 1984; Frayer, 1984). There is evidence for hybridization even where there may have been invasion in western Europe (Brauer, 1984; Stringer *et al.*, 1984). There is evidence for dental continuity from the present going at least as far back as the Mesolithic in Asia (Brace *et al.*, 1984). Morphological continuity does not mean that there were no invasions, incursions, or demic diffusions into those areas since the time of *Homo erectus*, but they do strongly suggest that (1) *in situ* racial variation may have an age much more than even a hundred thousand years, and (2) there was a considerable amount of admixture between the indigenes and any peoples diffusing into their territories. Of course, it must be acknowledged in regard to morphological continuity that consensus moves with rapid as well as unpredictable changes in direction as new material is found (sometimes even without it).

Conclusion

In sum, while there is a non-identifiability problem which prevents us arriving at a definitive reconstruction of the major human races, it is this author's view that the migrational implications of recent separation times make such times difficult to accept. Even if today's major races are descended from some local ancestral source population, subsequent gene flow amongst human populations would have led to a substantial decrease in the separation time estimated from current gene frequency data. And a simple separation/bifurcation model may itself not be realistic. It may be more likely that modern *Homo sapiens* evolved as a single species in the Old World, but has differentiated locally over many thousands of years.

A la recherche des tombes perdues

In searching for evidence of the processes of migration in the human past, it is important to remember that we are relying only on the tombs we have found. Those tombs consist of skeletons and, in a way, of the genes available for sampling. The latter are the present-day biological remains of their bearers' ancestors. But these various remains do not represent all who lived in times past. There are many lost tombs. Those that are found may not be sufficient for us to draw a very accurate picture of the events that preceded us. Where this *is* the case, statistical descriptions of the pattern of differences among living peoples are informative and surely have at least some relevance to history. However, it would be preferable not to use such incomplete data to make reconstructions which imply ecological or behavioural capabilities of our ancestors without at least offering plausible support for them.

It is obvious that gene flow has always been important in maintaining the continuity of hominid species and in modifying their patterns of diversity. It also seems clear that the tempo of migrational events has greatly accelerated since the evolution of human culture. One result is that much of current human diversity is directly traceable to recent large-scale phenomena. What is still not known, however, is the degree to which systematic demic diffusion among pre-agricultural peoples, or at the *sapiens–erectus* border, was a factor in creating the pattern of that diversity.

ACKNOWLEDGMENTS. This paper has benefited from the work of several colleagues and friends. I wish particularly to acknowledge the enlightenment as well as the extraordinary assistance I have received from Dr Peter E. Smouse in the preparation of this paper. His influence on my thinking is interwoven throughout the manuscript. I also thank Dr Emoke J. E. Szathmary for the substantial help she provided regarding the genetics of North American peoples, and for her typically thorough critique of the entire manuscript. Of course, it cannot be assumed that these two will smile approvingly at the outcome. Similarly, although I differ with some of his interpretations, our immense debt to L. L. Cavalli-Sforza should be obvious.

References

Adovasio, J. M., Donahue, J., Cushman, K., Carlisle, R. C., Stuckenrath, R., Gunn, J. D. & Johnson, W. C. (1983). Evidence from Meadowcroft Rockshelter. In *Early Man in the New World*, ed. R. Shutler, pp. 163–81. Beverley Hills, Cal: Sage Publications.

Ammerman, A. J. & Cavalli-Sforza, L. L. (1985). *The Neolithic Transition and the Genetics of Populations in Europe*. Princeton: Princeton University Press.

Bellwood, P. (1975). The prehistory of Oceania. *Current Anthropology*, 16, 9–28.

Bodmer, W. F. & Cavalli-Sforza, L. L. (1976). *Genetics, Evolution and Man*. San Francisco: W. Freeman.

Brace, C. L., Saho, X. & Zhang, Z. (1984). Prehistoric and modern tooth size in China. In *The Origins of Modern Humans: A World Survey of the Fossil Evidence*, ed. F. H. Smith & F. Spencer, pp. 485–516. New York: A. R. Liss.

Brauer, G. (1984). A craniological approach to the origin of anatomically modern *Homo sapiens* in Africa and implications for the appearance of modern Europeans. In *The Origins of Modern Humans: A World Survey of the Fossil Evidence*, ed. F. H. Smith & F. Spencer, pp. 327–410. New York: A. R. Liss.

Brown, W. M. (1980). Polymorphism in mitochondrial DNA of humans revealed by restriction endonuclease analysis. *Proceedings of the National Academy of Sciences, USA*, 77, 3605–9.

Cann, R. L. (1982). *The Evolution of Human Mitochondrial DNA*. Ph.D. Thesis, University of California, Berkeley.

Cann, R. L. & Wilson, A. C. (1983). Length mutations in human mitochondrial DNA. *Genetics*, 104, 699–711.

Cavalli-Sforza, L. L. (1969). Human diversity. *Proceedings, XII International Congress of Genetics, Tokyo*, 3, 405–16.

Cavalli-Sforza, L. L. (1971). Pygmies, an example of hunters–gatherers, and genetic consequences for man of domestication of plants and animals. *Human Genetics, Proceedings of the Fourth International Congress of Human Genetics*, (Amsterdam: Excerpta Medica), pp. 79–95.

Cavalli-Sforza, L. L. (1983). Isolation by distance. In *Human Population Genetics: The Pittsburgh Symposium*, ed. A. Chakravarti, pp. 229–48. New York: Van Nostrand Reinhold).

Cavalli-Sforza, L. L. & Bodmer, W. F. (1971). *The Genetics of Human Populations*. San Francisco: W. Freeman.

Chagnon, N. (1968). *Yanomama: the Fierce People*. New York: Holt, Rinehart, and Winston.

Chakraborty, R., Blanco, R., Rothhammer, F. & Llop, E. (1976). Genetic

162   K. M. Weiss

variability in Chilean Indian populations and its association with geography, language, and culture. *Social Biology*, 23, 73–82.

Chapman, R. (1985). The seeds of agriculture (review). *Nature*, 317, 391.

Colinvaux, P. A. (1985). Land bridge of Duvanny Yar, *Nature*, 314, 581–2.

Colinvaux, P. A. & West, F. H. (1984). The Beringian ecosystem. *Quarterly Review of Archaeology*, 5, 10–15.

Crow, J. F. & Kimura, M. (1970). *An Introduction to Population Genetics Theory*. New York: Harper and Row.

Denaro, M., Blanc, H., Johnson, M. J., Chen, K. H., Wilmsen, E., Cavalli-Sforza, L. L. & Wallace, D. C. (1981). Ethnic variation in *Hpa* I endonuclease cleavage patterns of human mitochondrial DNA. *Proceedings of the National Academy of Sciences of the US*, 78, 5768–72.

Fladmark, K. R. (1983). Times, and places: environmental correlates of Mid-to-Late Wisconsinan human population expansion in North America. In *Early Man in the New World*, ed. R. Shutler, pp. 13–41. Beverley Hills, Cal.: Sage Publications.

Frayer, D. W. (1984). Biological and cultural change in the European Late Pleistocene and Early Holocene. In *The Origins of Modern Humans: A World Survey of the Fossil Evidence*, ed. F. H. Smith & F. Spencer, pp. 211–50. New York: A. R. Liss.

Gould, S. J. (1982). Darwinism and the expansion of evolutionary theory. *Science*, 216, 380–7.

Griffin, J. B. (1979). The origin and dispersion of American Indians in North America. In *The First Americans: Origins, Affinities, and Adaptations*, ed. W. S. Laughlin & A. B. Harper, pp. 43–56. New York: Gustav Fischer.

Guglielmino-Matessi, C. R., Gluckman, P. & Cavalli-Sforza, L. L. (1979). Climate and the evolution of skull metrics in man. *American Journal of Physical Anthropology*, 50, 549–64.

Habgood, P. J. (1985). The origin of the Australian Aborigines: an alternative approach and view. In *Hominid Evolution: Past, Present, and Future*, ed. P. V. Tobias, pp. 367–80. New York: A. R. Liss.

Harpending, H. C. & Jenkens, T. (1974). !Kung population structure. In *Genetic Distance*, ed. J. F. Crow, pp. 137–61. New York: Plenum.

Harper, A. B. (1980). Origins and divergence of Aleuts, Eskimos and American Indians. *Annals of Human Biology*, 7, 547–54.

Hill, A. V. S., Bowden, D. K., Flint, J., Whitehouse, D. B., Hopkinson, D. A., Oppenheimer, S. J., Serjeantson, S. W. & Clegg, J. B. (1986). A population genetic survey of the haptoglobin polymorphism in Melanesians by DNA analysis. *American Journal of Human Genetics*, 38, 382–9.

Hill, A. V. S., Bowden, D. K., Trent, R. J., Higgs, D. R., Oppenheimer, S. J., Thein, S. L., Mickleson, K. N. P., Weatherall, D. J. & Clegg, J. B. (1985). Melanesiand and Polynesians share a unique α-Thalassemia mutation. *American Journal of Human Genetics*, 37, 571–80.

Howell, N. (1976). Toward a uniformitarian theory of human paleodemography. In *The Demographic Evolution of Human Populations*, ed. R. H. Ward & K. M. Weiss, pp. 25–40. London: Academic Press.

Howells, W. W. (1973a). *The Pacific Islanders*. New York: Scribner's.

Howells, W. W. (1973b). *Cranial Variation in Man*. Papers of the Peabody

Museum of Archaeology and Ethnology **67**, 259 pp. Cambridge, Mass.: Harvard University.

Howells, W. W. (1980). *Homo erectus* – who, when and where: a survey. *Yearbook of Physical Anthropology*, **23**, 1–24.

Johnson, M. J., Wallace, D. C., Ferris, S. D., Rattazzi, M. C. & Cavalli-Sforza, L. L. (1983). Radiation of human mitochondria DNA types analyzed by restriction endonuclease cleavage patterns. *Journal of Molecular Evolution*, **19**, 255–71.

Jones, J. S. & Rouhani, S. (1986). How small was the bottleneck? *Nature*, **319**, 449–50.

Jorde, L. B. (1980). The genetic structure of subdivided human populations: a review. In *Current Developments in Anthropological Genetics. I. Theory and Methods*, ed. J. H. Mielke & M. H. Crawford, pp. 135–208. New York: Plenum.

Kirk, R. L. (1979). Genetic differentiation in Australia and the Western Pacific and its bearing on the origin of the first Americans. In *The First Americans: Origins, Affinities, and Adaptations*, ed. W. S. Laughlin & A. B. Harper, pp. 211–37. New York: Gustav Fischer.

Kirk, R. L. & Thorne, A. G. (eds) (1976). *The Origin of the Australians*. Canberra: Australian Institute of Aboriginal Studies.

Latter, B. D. H. (1973). Measures of genetic distance between individuals and populations. In *Genetic Structure of Populations*, ed. N. E. Morton, pp. 27–39. Honolulu: University of Hawaii Press.

Laughlin, W. S., Jorgensen, J. B. & Frohlich, B. (1979). Aleuts and Eskimos: survivors of the Bering land bridge coast. In *The First Americans: Origins, Affinities, and Adaptations*, ed. W. S. Laughlin & A. B. Harper, pp. 91–104. New York: Gustav Fischer.

Livingstone, F. B. (1980). Natural selection and random variation in human evolution. In *Current Developments in Anthropological Genetics. I. Theory and Methods*, ed. J. H. Mielke & M. H. Crawford pp. 87–110. New York: Plenum.

Lynch, T. F., Gillespie, R., Gowlett, J. A. J. & Hedges, R. E. M. (1985). Chronology of Guitarrero Cave, Peru. *Science*, **229**, 864–7.

May, R. M. (1984). Prehistory of Amazonian Indians. *Nature*, **321**, 19–20.

Menozzi, P., Piazza, A. & Cavalli-Sforza, L. L. (1978). Synthetic maps of human gene frequencies in Europe. *Science*, **201**, 786–92.

Murdock, G. P. (1934). *Our Primitive Contemporaries*. New York: Macmillan.

Neel, J. V. (1971). 'Lessons' from a primitive people. *Science*, **170**, 815–22.

Neel, J. V. (1978). The population structure of an Amerindian tribe, the Yanomama. *Annual Review of Genetics*, **12**, 365–413.

Neel, J. V. (1984). The *real* human populations. In *Human Population Genetics: the Pittsburgh Symposium*, ed. A. Chakravarti, pp. 249–73. New York: Van Nostrand Reinhold.

Nei, M. (1978). The theory of genetic distance and the evolution of the human races. *Japanese Journal of Human Genetics*, **23**, 341–69.

Nei, M. (1982). Evolution of human races at the gene level. In *Human Genetics. Part A: The Unfolding Genome*, ed. Bonne-Tamir, pp. 167–81. New York: A. R. Liss.

Nei, M. (1985). Human evolution at the molecular level. In *Proceedings, Oji Conference on Population Genetics and Evolution*, (in press).

Nei, M. & Roychoudhury, A. K. (1974). Genetic variation within and between the three major races of man, Caucasoids, Negroids, and Mongoloids. *American Journal of Human Genetics*, **26**, 421–43.

Nei, M. & Roychoudhury, A. K. (1982). Genetic relationship and evolution of human races. In *Evolutionary Biology*, vol. 14, ed. M. K. Hecht, B. Wallace & C. T. Prince, pp. 1–59. New York: Plenum.

O'Rourke, D. H., Suarez, B. K. & Crouse, J. D. (1985). Genetic variation in North Amerindian populations: covariance with climate. *American Journal of Physical Anthropology*, **67**, 241–50.

Owen, R. C. (1984). The Americas: the case against an ice-age human population. In *The Origins of Modern Humans: A World Survey of the Fossil Evidence*, ed. F. H. Smith & F. Spencer, pp. 517–63. New York: A. R. Liss.

Piazza, A., Sgarmella-Zonta, L., Gluckman, P. & Cavalli-Sforza, L. L. (1975). The Fifth Histocompatibility Workshop gene frequency data: a phylogenetic analysis. *Tissue Antigen*, **5**, 445–63.

Piazza, A. & Cavalli-Sforza, L. L. (1983). Treeness tests and the problem of variable evolutionary rates. In *Numerical Taxonomy*, ed. J. Felsenstein, pp. 451–63. Berlin: Springer.

Piazza, A., Menozzi, P. & Cavalli-Sforza, L. L. (1981). Synthetic gene frequency maps of man and selective effects of climate. *Proceedings of the National Academy of Sciences of the USA*, **78**, 2638–42.

Rightmire, G. P. (1981). Later Pleistocene hominids of eastern and southern Africa. *Anthropologie*, **19(1)**, 15–26.

Rightmire, G. P. (1984). *Homo sapiens* in Sub-Saharan Africa. In *The Origins of Modern Humans: A World Survey of the Fossil Evidence*, ed. F. H. Smith & F. Spencer, pp. 295–325. New York: A. R. Liss.

Roychoudhury, A. (1984). Genetic relationship between Indian tribes and Australian aboriginals. *Human Heredity*, **34**, 314–20.

Roychoudhury, A. & Nei, M. (1985). Genetic relationship between Indians and their neighboring populations. *Human Heredity*, **35**, 201–6.

Smith, F. H. (1984). Fossil hominids from the Upper Pleistocene of Central Europe and the origin of modern Europeans. In *The Origins of Modern Humans: A World Survey of the Fossil Evidence*, ed. F. H. Smith & F. Spencer, pp. 137–209. New York: A. R. Liss.

Smouse, P. E. (1983). Genetic architecture of swidden agricultural tribes from the lowland rain forests of South America. In *Current Developments in Anthropological Genetics. 2. Ecology and Population Structure*, ed. M. H. Crawford & J. H. Mielke, pp. 139–78. New York: Plenum.

Sokal, R. R. & Menozzi, P. (1982). Spatial autocorrelations of *HLA* frequencies in Europe support demic diffusion of early farmers. *American Naturalist*, **119**, 1–17.

Spielman, R. S., Migliazza, E. C. & Neel, J. V. (1974). Regional linguistic and genetic differences among Yanomama Indians. *Science*, **184**, 637–44.

Spielman, R. S. & Smouse, P. E. (1976). Multivariate classification of human populations. I. Allocation of Yanomama Indians to villages. *American Journal of Human Genetics*, **28**, 317–31.

Spuhler, J. N. (1972). Genetic, linguistic, and geographical distances in native North America. In *The Assessment of Population Affinities in Man*, ed. J. Weiner & J. Huizinga, pp. 72–95. Oxford: Clarendon.

Spuhler, J. N. (1979). Genetic distances, trees, and maps of North American Indians. In *The First Americans: Origins, Affinities, and Adaptations*, ed. W. S. Laughlin & A. B. Harper, pp. 135–83. New York: Gustav Fischer.

Stringer, C. B., Hublin, J. J. & Vandermeersch, B. (1984). The origin of anatomically modern humans in Western Europe. In *The Origins of Modern Humans: A World Survey of the Fossil Evidence*, ed. F. H. Smith & F. Spencer, pp. 51–135. New York: A. R. Liss.

Suarez, B. K., Crouse, J. D. & O'Rourke, D. H. (1985). Genetic variation in North Amerindian populations: the geography of gene frequencies. *American Journal of Physical Anthropology*, **67**, 217–32.

Sukernik, R. I. & Osipova, P. (1982). Gm and Km immunoglobulin allotypes in Reindeer Chukchi and Siberian Eskimos. *Human Genetics*, **61**, 148–53.

Szathmary, E. J. E. (1979a). Blood groups of Siberians, Eskimos Subarctic and Northwest Coast Indians: the problem of origins and genetic relationships. In *The First Americans: Origins, Affinities, and Adaptations*, ed. W. S. Laughlin & A. B. Harper, pp. 185–209. New York: Gustav Fischer.

Szathmary, E. J. E. (1979b). Eskimo and Indian contact: examination of craniometric, anthropometric and genetic evidence. *Arctic Anthropology*, **16**(2), 23–46.

Szathmary, E. J. E. (1981). Genetic markers in Siberian and northern North American populations. *Yearbook of Physical Anthropology*, **24**, 37–73.

Szathmary, E. J. E. (1984). Peopling of northern North America: clues from genetic studies. *Acta Anthropologica*, **8**, 79–110.

Szathmary, E. J. E., Ferrell, R. E. & Gershowitz, H. (1983). Genetic differentiation in Dogrib Indians: serum protein and erythrocyte enzyme variation. *American Journal of Physical Anthropology*, **62**, 249–54.

Szathmary, E. J. E. & Ossenberg, N. S. (1978). Are the biological differences between North American Indians and Eskimos truly profound? *Current Anthropology*, **19**, 673–701.

Trinkaus, E. (1984). Western Asia. In *The Origins of Modern Humans: A World Survey of the Fossil Evidence*, ed. F. H. Smith & F. Spencer, pp. 251–93. New York: A. R. Liss.

Turner, C. G. (1983). Dental evidence for the peopling of the Americas. In *Early Man in the New World*, ed. R. Shutler, pp. 147–57. Beverley Hills, Cal.: Sage Publications.

Wainscoat, J. S., Hill, A. V. S., Boyce, A. L. [sic – should be 'J'], Flint, J., Hernandez, M., Thein, S. L., Old, J. M., Lynch, J. R., Falusi, A. G., Weatherall, D. L. & Clegg, J. B. (1986). Evolutionary relationships of human populations from an analysis of nuclear DNA polymorphisms. *Nature*, **319**, 491–3.

Ward, R. H. (1976). *Order and Chaos: genetic structure and social systems in a tribal population.* Occasional Papers, pp. 1–58. Mexico City: Instituto de Investigaciones Antropologicas, Universidad Nacional Autonoma de Mexico.

Weiss, K. M. (1984). On the number of members of the genus *Homo* who have

166   K. M. Weiss

ever lived, and some evolutionary implications. *Human Biology*, **56**, 637–50.
Weiss, K. M. & Maruyama, T. (1976). Archeology, population genetics and studies of human racial ancestry. *American Journal of Physical Anthropology*, **44**, 31–50.
Wijsman, E. M. & Cavalli-Sforza, L. L. (1984). Migration and genetic population structure with special reference to humans. *Annual Review of Ecology and Systematics*, **15**, 279–301.
Williams, R. C., Steinberg, A. G., Gershowitz, H., Bennett, P. H., Knowler, W. C., Pettitt, D. J., Butler, W., Baird, R., Dowda-Rea, L., Burch, T. A., Morse, H. G. & Smith, C. G. (1985). Gm allotypes in Native Americans: evidence for three distinct migrations across the Bering Land Bridge. *American Journal of Physical Anthropology*, **66**, 1–20.
Wolpoff, M. H. (1980). *Paleoanthropology*. New York: Knopf.
Wolpoff, M. H. (1985). Human evolution at the peripheries: the pattern at the eastern edge. In *Human Evolution: Past, Present, and Future*, ed. P. V. Tobias, pp. 355–65. New York: A. R. Liss.
Wolpoff, M. H., Wu, W. X. & Thorne, A. G. (1984). Modern *Homo sapiens* origins: a general theory of hominid evolution involving the fossil evidence from East Asia. In *The Origins of Modern Humans: A World Survey of the Fossil Evidence*, ed. F. H. Smith & F. Spencer, pp. 411–83. New York: A. R. Liss.

7 Migration and adaptation

MICHAEL A. LITTLE AND PAUL T. BAKER

Introduction

Adaptation to the environment

Migration or movement as a means of behavioural adaptation to environmental stress is a fundamental property of all mobile forms of life. Human populations are characterized as highly mobile, not only because of their long palaeolithic existence as nomadic hunters and gatherers, but also because of the worldwide movements following the age of exploration and the widespread migration occurring today (Davis, 1974; Clarke, 1984). The effects of these population movements on populations of origin, migrants themselves, host populations, and each of their respective environments are complex and still not fully understood (Lee, 1966; Graves & Graves, 1974; Robins, 1981; Baker & Baker, 1984). We do know that although humans' motives for moving are usually to improve their lives or the lives of their children, the new environments to which they are exposed are often stressful and carry high risks of illness and death. The reasons for these high risks are manifold, but are based fundamentally on the migrants' lack of adaptation to the new environment.

Within the framework of evolutionary and adaptation theory, there are a number of general principles that will assist in the understanding of the biological effects of migration. First, populations with relatively stable social and subsistence systems and which have been long resident in a given area have adapted in their biology, population structure and sociocultural system to this area and its environment. The basis for these patterns is 'selection', that is selection for compatible genetic, plastic, social and behavioural adaptations (Baker, 1966, 1984b; Skinner, 1981; Little, 1982). All biological adaptive patterns have genetic bases; however, some are fairly specific and easily identifiable from their phenotypic expression, while others are more plastic or flexible and are expressed more broadly depending on environmental circumstances. Short-term and developmental adaptations fall into this latter category. Some popu-

167

lations are better adapted than others to given environments and environmental stress by virtue of their genetic makeup (Baker, 1966). This has been demonstrated many times (Harrison *et al.*, 1977; Ortner, 1983). Second, populations which undergo change will be moved to a state of stress or instability until new modes of adaptation occur (Baker, 1977c). Several hypotheses have been suggested (Baker, 1977c; 1987) to define further the biological effects of migration:

(1) The more rapid and dramatic changes will produce greater stress as manifested by illness, loss of biological fitness and declines in sociocultural integration than will gradual and minor changes.

(2) Adults, particularly the elderly, will suffer more from environmental change than will children. This is based on the knowledge that some adaptations require development during the period of child and adolescent growth.

(3) Effects of environmental change on health and state of adaptation will be most acute immediately after a move. Adjustments will occur as exposure to the new environment increases.

There is some evidence to support each of these three hypotheses yet they are not quite at the level of general principles.

Changing environments and stress

Until the 'Neolithic and Urban Revolutions' between 12 000 and 6000 years ago (Childe, 1952), environmental change in human populations was very gradual. The rise of civilization produced a new sociopolitical form of organization that led to increased social stratification, occupational specialization, mobility through trade and conquest, nucleated settlement, food surplus, and new disease patterns. These conditions stimulated diversity and change that has continued to the present. Within the past 200 years, change has accelerated at three dimensional levels. *Culture change* has flowed from Western European industrial development to non-Western areas of the world in a process referred to as 'Westernization'. As Western cultures evolved, *time changes* occurred as the process of change ('secular trend') was manifested in increases in size of children at all ages, earlier maturation, and increased adult size (Eveleth & Tanner, 1976, pp. 260–1). Migration, of course, led to *spatial changes* in both the human social and the natural attributes of the environment. Modernization in the West was a gradual process of change. Modernization of traditional, peasant or tribal societies and migration to modernized areas have both led to more rapid change and stress associated with rapid environmental change. Present programmes of 'economic development' of peasant and tribal peoples also contribute to rapid processes of modernization and Westernization (Scudder, 1980; Moran, 1981).

An important point to be made is that the pattern of culture change and its biocultural effects produced by Westernization, modernization or migration is very much the same. There are, of course, some differences. *In situ* changes associated with economic, political and educational directions toward modernization do not involve exposure to a new physical environment; residential changes associated with migration do, which may intensify the constellation of stresses imposed on the migrants. Finally, the city, which may be viewed as the epitome of a modernized residence, will be the major environment for most of the world's population in the future. Cities are also the primary receiving areas for migrants and their high population growth rates reflect this phenomenon. Throughout the world, there were only 75 cities with populations greater than one million in the year 1950; in 1985, it is estimated that there are nearly 275 cities with populations this size or greater (Briggs, 1983). In identifying the effects of migration on the biology of human populations, the urban environment must be considered carefully (Harrison & Gibson, 1976).

Research designs

As noted below, Boas (1912) conducted the first study to test a hypothesis in which migrant populations were a part of the design. Boas' design was a simple but effective one (see Fig. 7.1(*a*)). Sedente adults and children in Europe were compared with first and second generation migrants in New York City. The changes in head shape and stature that Boas observed demonstrated that these structures showed a high degree of genetic plasticity. Many migration studies followed Boas' lead in attempts to define which anthropometric measures were relatively inflexible in their phenotypic expression and, hence, under tight genetic control, and which were not (Shapiro, 1939; Goldstein, 1943; Lasker, 1946).

Later investigators were concerned with the question of selective migration or pre-selection of migrants for specific biological traits that would appear to have been caused by exposure to the new environment. In early tests, Lasker (1952, 1954) compared anthropometric measures of Mexican migrants who moved as children or as adults to the United States with Mexican sedentes; he also compared brothers who migrated at different ages or who did not migrate. He found no evidence for selective migration for physical measures of size. In a more recent study, Malina

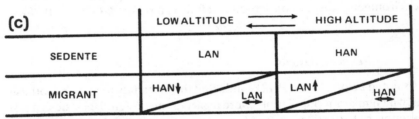

Fig. 7.1. Research designs for studies of sedente and migrant populations. (*a*) Boas' (1912) study of European migrants to New York City; (*b*) Harrison's (1966) design to study altitude effects by use of upward and downward migrants; (*c*) Baker's (1976) expansion of Harrison's (1966) design.

and his co-workers (1982) did follow-up measurements on migrants from Oaxaca to Mexico City who had been measured before migration. This study supported Lasker's findings. Susanne (1984) reviewed this work and confirmed the lack of evidence for selective migration. However, one study did find a greater stature in migrants who moved within the United Kingdom *vs.* non-migrants (Mascie-Taylor, 1984). A question that has been raised less frequently is that of kin selection and its effects within migrant streams. Leslie (1985) provided ample evidence for kin-based selection among migrants from behavioural and sociocultural data and on theoretical grounds, but acknowledged that its phenotypic expression is difficult to demonstrate.

A basic framework to deal with the effects of heredity and a high-altitude environment was presented by Harrison (1966) for work in the Himalayas. It is illustrated in Fig. 7.1(*b*). The research framework added to Boas' (1912) original design by identifying a specific environmental variable – altitude – and by dealing with migrants both to and from high altitude. Baker (1976) discussed this framework as an example of what he called a 'multiple stress–multiple population' design where the four populations in Fig. 7.1(*b*) could be compared in six ways to test for different relationships. A modified form of this design was used by Baker (1976; Baker & Beall, 1982) in the research of downward migrants from the Peruvian Andes to the coast (see Fig. 7.1(*c*)). Migrants within the high-altitude zone and migrants within the low-altitude zone were used to control for effects of change of residence at both elevations. Although not illustrated here, 'intermediate altitude' was added to refine the design, and age at migration was a part of the design also. Haas (1976) employed a non-migrant design to test for high-altitude effects, where he included categories of urban–rural residence and ethnicity. Inclusion of these treatments in designs such as that represented in Fig. 7.1(*c*) carries with it typical problems of small sample sizes for each of the design cells.

A final design can be used to illustrate studies of migration in which there are major changes in subsistence and life style. Fig. 7.2 is a simple design for comparing East African traditional pastoralists with former pastoralists who have changed their means of subsistence either in moving out of the region or in remaining at their home. Alternative patterns of subsistence and variations in the degree of sociocultural deviation from the traditional pastoral lifestyle impose stresses in nutri-

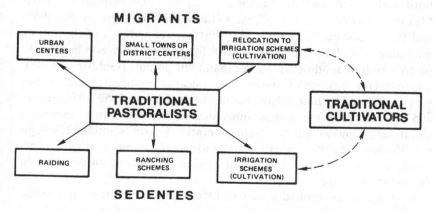

Fig. 7.2. A research framework to compare migrants and sedentes who have changed subsistence from traditional pastoralism to other means (solid arrows) with traditional cultivators (interrupted arrows). (Modified from Little, 1980.)

tion, disease, and general health on the population that does not migrate. Migration then superimposes other environmental stresses, the effects of which can be sorted out by comparisons of the migrants and the sedentes.

The scope of research in which migrants have and are able to be employed is substantial. In some ways migrants provide the richest series of 'natural experiments' in which humans can be studied without intrusion, intervention or manipulation. In the sections that follow, some of the findings of these natural experiments will be discussed within the framework of human adaptation to the environment.

Patterns of stress and adaptation

Growth and adult size

The earliest, and perhaps, best known study of migrants was conducted by Franz Boas (1912) to determine if the American-born children of European immigrants were different from their parents. In this study, designed to challenge the prevailing dogma at that time of 'the fixity of human types or races', changes in head dimensions and slight increases in stature characterized those born in New York in contrast to their European-born parents. This was, of course, a clear demonstration of developmental plasticity or adaptation during the child and adolescent period of growth and development (Lasker, 1969). Many anthropologists followed Boas' in using the migration model to demonstrate and quantify the effects of environmental changes on anthropometric measures in native-born and American-born Japanese (Shapiro, 1939; Greulich, 1957), Mexicans (Goldstein, 1943), Chinese (Lasker, 1946), Swiss (Hulse, 1957) and Puerto Ricans (Thieme, 1957). Further refinements of the simple migration model attempted to control for the secular trend in growth, where children as adults are larger than their parents (Goldstein, 1943), and to test for selection of migrants in physical or biological characteristics when compared to their sedente counterparts (Lasker, 1952, 1954; Hulse, 1957). Results of these studies almost uniformly demonstrated increased size in height and weight of second generation migrants, minimal biological differences between migrants and sedentes (minimal or no pre-selection), and that second generation migrants reflected an accelerated secular trend in maturation.

Several studies applied a variant form of the migrant design to the growth of United States children who were resident abroad. Mills (1942) measured height and weight of American children who were either born and reared in the United States Panama Canal Zone, or born in the United States and resident in the Zone for less than a year. Height differences for boys and girls were slight, while Panama-born children were more linear (low weight/height ratios) at all ages than the children born in the United States. Eveleth (1966) conducted a similar study in Rio de Janeiro of longitudinal growth of upper middle class American children who had been resident in Rio for about half their lives. They were compared with United States growth data where the American migrant children in Rio showed a tendency toward linearity that persisted between ages 6 and 16 years. Both Mills and Eveleth attributed this tendency toward linearity to tropical residence and heat effects. Although these children in the Canal Zone and Rio were probably on diets comparable to those of their United States counterparts, physical activity levels of the various groups of children were not controlled. In a third study of this nature, Johnston and co-workers (1976) compared longitudinal growth of well-off, Westernized children attending the American School in Guatemala City. American and European children were compared with native Guatemalan children, and each was compared with longitudinal growth data from a California study. The American and European migrant children grew more like Guatemalan children before adolescence and more like California children during adolescence. The authors suggested that pre-adolescent growth is more sensitive to environmental influences than is adolescent growth.

A number of studies have been conducted of Chinese, Japanese and Korean children that represent current interests in growth and migration. Kondo & Eto (1975) conducted cross-sectional studies of American-born Japanese in Los Angeles and Japanese in Nagoya city and found that the differences in height and weight were insignificant. Apparently, the secular trend of growth in Japan produced increases in size of children that was equivalent to the short-term changes experienced by second generation migrants in the United States (Kondo & Kobayashi, 1975). This was not the case when London Chinese toddlers ($\leq$ 5 yr) were compared with Hong Kong age-mates (Wheeler & Tan, 1983). Heights of the London Chinese children were slightly less than London native norms and slightly greater than Hong Kong Chinese. Weight ranking was in the same order, but after two years of age the differences between the three groups were marked. When Korean school children who were born and raised in Japan were compared with Japanese children in Japan (relatively tall) and Korean children in Korea (relatively short), the second generation migrant Korean children were taller than both sedente groups (Kim, 1982).

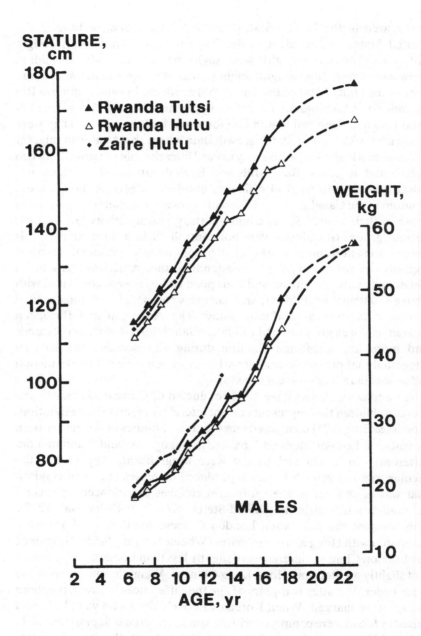

Fig. 7.3. Comparisons of the growth in stature and weight of sedentary Rwanda Tutsi and Hutu with Hutu children whose families had migrated to Zaïre. (Modified from Hiernaux, 1964.)

One of the few studies of migration and growth under non-Western conditions was carried out by Hiernaux (1964) on Rwanda Tutsi pastoralists and Hutu cultivators and a population of migrant Hutu who were transported from Rwanda to Katanga Province in Zaïre to work in the copper mines. In Rwanda, the dominant Tutsi and subordinate Hutu were bound by strong socioeconomic ties, and the general living conditions of the sedente Hutu were inferior to the Tutsi. Living conditions for the migrant Hutu in Katanga included improved hygiene, diet and medical care. Within this framework Hiernaux measured height and weight of the three groups: Tutsi and Hutu sedentes aged 6–17 years and Hutu migrants aged 6–13 years. Fig. 7.3 indicates that Hutu migrants achieved intermediate stature but highest weight levels. The relatively enriched Hutu migrant environment appeared to enhance growth in size, particularly weight. The less developmentally adaptable variable of stature in the Hutu did not reach Tutsi levels.

With a few exceptions, as in the case of Hiernaux' (1964) design, and as Beall (1982) has observed, the fundamental Boasian (1912) design of migrant studies has changed very little in the past 75 years. The basic weakness of the design is the difficulty in controlling for environmental complexity. When more rigorous environmental controls and longitudinal studies have been applied, the design is likely to continue to provide more information on child and adolescent growth.

Temperature regulation and metabolism

Migrant populations have been used relatively infrequently in studies of human adaptation to environmental temperature extremes. Most studies were designed to test the effects of a given environment on temperature regulation, either through an imposed heat or cold stress, or by tests of subjects in a resting state, as in basal metabolic rate (BMR). A few studies have attempted to sort out hereditary, developmental and short-term adaptations at physiological and morphological levels in the body with varying degrees of success (cf. Frisancho, 1979; Roberts, 1978).

Humans, as homeotherms, maintain body temperatures within a fixed range by continually producing heat (as a by-product of energy metabolism) and by dissipating the excess, via the surface of the body, to the environment. Basal heat production or BMR is a variable that is inversely correlated at the population level with environmental temperature (Roberts, 1952, 1978). BMR at the individual level is a function of age, sex, size, body composition, physiological state and state of thermal adaptation (Ganong, 1981; Hong, 1963). Based on the knowledge that a low BMR would be useful in the tropics, studies were conducted to determine if sojourners or migrants would show depressed BMRs while residing in tropical conditions. In India, Malhotra, Ramaswamy & Ray (1960) demonstrated that Indian BMRs were about 10% lower than those of expatriate Europeans. In an excellent research design, Mason & Jacob (1972) compared BMRs of natives of India and Europeans who lived and travelled between Bombay and London: both groups were tested in each location. The general pattern was one in which BMRs were elevated in London and depressed in Bombay on average about 5% for natives of both zones, and within two to six months of the move. The London/Bombay study, which demonstrated the plasticity of BMR, was extended by the work of Vallery-Masson, Bourlier & Poitrenaud (1980), who tested BMR in European expatriates who were long-term residents in West Africa and Indochina, but who had returned to France to retire. They were compared with an age-matched sample of French who had never been out of the country. Despite the fact that the expatriates had been back in France for an average of 10 to 12 years, their BMRs were about 8% lower than the controls. Among those who had been born in the tropics, their BMRs were about 13 to 17% lower than French controls. Body weights of former tropical residents and controls were equivalent although body composition was not known. The authors provided no explanation for these long-range effects.

Cold adaptation studies in which migrants were a part of the design have centred on Asian populations and extremity responses to acute cold. Yoshimura & Iida (1952) reported on four populations (Japanese, Chinese, Manchurian Mongols and native Orochon nomads) who were tested during an ice-water finger immersion during World War II. Orochons maintained warmest fingers, attributed by the authors to a lifetime of cold exposure or developmental adaptation. Japanese migrant adults did poorly on the test, but their children did as well as the Orochons, which supported the developmental adaptation hypothesis. Later studies by Yoshimura and his associates (Yoshimura & Horvath, 1975; Hori, Taguchi & Horvath, 1975) of Japanese migrants to California have provided no new information. Eagan (1967) demonstrated the persistence of cold adaptation by testing Alaskan Eskimos who had been resident in Oregon for nine months. In an acute finger-cooling test, they were able to maintain their characteristically warm skin temperatures when contrasted with experienced non-Eskimo mountain climbers. Little et al. (1971) also found a persistent ability of lowland Quechua Indians to maintain warm hand temperatures despite a long residence at temperate to subtropical sea level. Chinese of northern and southern China ancestry and who had been resident in Buffalo, New York, for three years were tested during hand immersion in 5°C water (So, 1975). So (1975) attributed the warmer responses of the north Chinese as evidence of a persistent genetic adaptation. Evidence against a genetic interpretation was provided by Steegmann (1974) who tested hand cooling in Hawaiian-born subjects of European and Japanese ancestry whose responses were the same.

The research designs employing migrants have provided some information to assist in sorting out genetic, developmental and short-term adaptation effects, but have not provided definitive answers.

Blood pressure and hypertension

One of the most distinct effects of migration, or any environmental change imposed on a population, is an increase in blood pressure. Huizinga (1972) compiled data on 87 populations in which 76 were non-European, and compared regression coefficients of systolic and diastolic blood pressure on age. He found that all European populations displayed increases in blood pressure with age, but that 27 of the 75 non-European populations showed no statistically significant age changes in blood pressure. He concluded that non-Westernized populations are characterized by low blood pressures and that it is only under conditions of acculturation or modernization that blood pressures become elevated and age-related increases occur. Genetic factors in elevated blood pressure and hypertension are important within the population level in terms of familial patterns, associations with obesity, sodium sensitivity and the like (Siervogel, 1983). The effects of genetic factors at the interpopulation level are less clear. There appears to be some agreement that sociocultural and associated variables contribute to age increases in blood pressure and hypertension (Dressler, 1982; Ward, 1983). Ward (1983) suggested that there are four key factors that characterize traditional tribal societies with low blood pressure levels: low salt use, lean body build or physique, high levels of physical activity, and low levels of psychosocial stress. However, these four key factors are not always present in normotensive native populations (Baker & Crews,1987). A discussion of African migrant groups will consider some of these questions.

The forced relocation of West African populations to the New World during the slave-trade period up to about 1850 was an example of the largest migration of tribal peoples known (Crosby, 1973). Today, Afro-Americans in the United States have considerably higher systolic and diastolic blood pressures than Euro-Americans (cf. Roberts & Mauer, 1977; Gibson & Gibbons, 1982). When socioeconomic status is considered, low socioeconomic level Afro-Americans have the highest elevations of blood pressure (Roberts & Mauer, 1977), whereas in many parts of Africa it is the upper socioeconomic groups that display highest blood pressures (Parry, 1969; Shaper & Saxton, 1969). Vaughan & Miall (1979) compared rural black Jamaicans with Tanzanians and Gambians, also from rural communities, and found marked differences in blood pressures between the three groups. Jamaicans were highest, Tanzanians intermediate and Gambians lowest. The authors attributed the differences to economic development and modernization in Jamaica. In St Lucia, West Indies, hypertension is very prevalent among the present descendants of West African slaves (Dressler, 1982). Here, it was found that those individuals at greatest risk of hypertension were those who were striving to achieve a more Western life style, but who had few resources to do so.

Among African populations, it is rural-to-urban migration in which blood pressure elevations have been best documented: Serer pastoralists of Senegal (Beiser *et al.*, 1976) and the Zulu of South Africa (Scotch, 1960) are notable examples. An equivalent phenomenon has been recorded by Shaper and his colleagues (1969) who measured blood pressure in Samburu pastoralists, some of whom were serving in the King's African Rifles (KAR) infantry unit prior to Kenyan independence in 1966. All infantrymen were originally pastoral nomads from northern Kenya rural bushland and were stationed in the cities of Nairobi, Nakuru and Nanyuki. They found that there was a positive relationship between years of service in the KAR and systolic blood pressure. Body weight was greater in the infantrymen than in the pastoralists, but this variable did not show a relationship to years of KAR service.

Many pastoral nomadic populations are moving to settled irrigation schemes adjacent to major rivers in northern Kenya. On these nucleated schemes the people have access to more traded and cultivated foods, including salt; have less access to milk, meat and other sources of protein; experience reduced activity levels; and are subject to a more regulated and bureaucratic system of subsistence than when they were nomads. Fig. 7.4 shows blood pressure for two groups of Turkana natives from north-west Kenya. The differences between the settled and nomadic populations were most pronounced in youths and young adults. Since the irrigation schemes had only been in existence for 15 to 20 years, it is likely that the settled adults in their 40s and 50s did not show elevated blood pressures because most had spent the greater part of their lives as pastoral nomads (Little *et al.*, 1986).

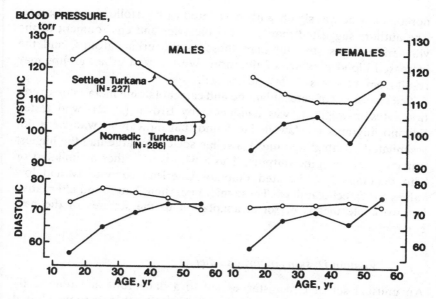

Fig. 7.4. Systolic and diastolic blood pressures of traditional nomadic Turkana pastoralists of Kenya with Turkana cultivators settled on an irrigation scheme (first generation migrants). (From Little *et al.*, 1987.)

Catecholamines and stress

Excretion of catecholamines (epinephrine and norepinephrine) is a good measure of sympathetic-adrenal medulary activity which is known to be a physiological response to stress (Levi, 1972; Ursin, Baade & Levine, 1978). Brown (1981) has argued that urinary catecholamine excretion should be used for studies of culture change and stress in human populations. There is an abundant literature identifying high levels of stress with the lifestyles of Western/urban society, including performance requirements (Carruthers, 1976; Jenner, Reynolds & Harrison, 1980; Reynolds *et al.*, 1981) and crowding (Lundberg, 1976), but little research has been conducted on migrants to date.

A study of natives from Western Samoa bears on the Polynesian migrant studies discussed below. James and his colleagues (1985) compared four group of young Samoan men in the independent state of Western Samoa. These were rural agriculturists, manual labourers, sedentary workers, and college students. The three latter groups were from the urban area of Apia. Villagers generally had lower catecholamine excretion rates and blood pressure than the three Apia groups. Also, norepinephrine was significantly correlated with systolic blood pressure. The authors suggested that '. . . the lifestyles and environment of the villagers are less stressful than those of the urban groups, and the increased blood pressure of the more Westernized urban groups may result from the stress of lifestyle change'.

The sole study of catecholamine and catecholamine metabolite excretion rates in migrants was conducted by Brown (1982), who tested Filipino migrants to Hawaii. He found that individuals who were not assimilated or integrated into Hawaiian society at large had the highest urinary catecholamine outputs. Those who were either assimilated or who maintained an isolated Filipino–American community had low values for catecholamines. These migrants, then, had found alternative but workable means of social adaptation to the stresses of the new environment.

Vitamin D, skin colour, and diet

An unusual set of circumstances led to a dangerous and potentially debilitating deficiency disease in migrants from South Asia to the United Kingdom. In the early 1970s, reports began accumulating of cases of rickets and osteomalacia among immigrants from India and Pakistan (Swan & Cooke, 1971; Chamberlain & Hosking, 1971; Clark, Simpson & Young, 1972). Serum assays of 25-hydroxycholecalciferol were used to estimate that roughly 25% of South Asian children were affected and that vitamin D deficiency was the major factor leading to the bone disorders (Preece et al., 1973). Later studies of larger samples of children from Glasgow (Goel et al., 1976) found up to 5% prevalence of florid rickets (with clinical signs) in South Asian children, but relatively low prevalence in African and Chinese immigrant children. In most of the surveys that were conducted in English and Scottish cities, the bulk of the cases of osteomalacia in pregnant women and rickets in children were South Asians. Africans and Afro-Caribbeans were affected, but to a lesser degree (Donovan, 1984).

The earliest reports (Clark et al., 1972) suggested that the deficiency diseases were due to inadequate intake of dietary vitamin D or a combination of this and a failure to synthesize cholecalciferol in the skin due to inadequate sunlight and heavy melanin pigmentation of the skin. South Asian populations have uniformly more melanin skin pigmentation than north-west Europeans (Kalla, 1983). However, an important dietary factor was identified by Ford and others (1972), who found that removing chapattis from the diet of several Pakistanis over a period of seven weeks produced dramatic improvements in their disease states. It was concluded that the high phytate content of the unleavened chapattis inhibited calcium mobilization and contributed to the likelihood that individuals consuming large amounts of chapattis would show signs of rickets (Ford *et al.*, 1972; Reinhold *et al.*, 1973). This epidemic of rickets appears to be controlled now by vitamin D fortification in South Asian children (Pietrek *et al.*, 1976; Goel *et al.*, 1981). The example of South Asian migrants in the United Kingdom is a particularly good case of a marked environmental change leading to a failure in adaptation – a failure that had cultural, dietary, genetic and environmental bases. It merits further discussion.

Rickets was relatively common among poor children who resided in the congested and polluted industrial cities of Western Europe during the last century (Loomis, 1970). Recent years have seen the gradual eradication of the disease in European urban children through a combination of improved living conditions, pollution control and vitamin D fortification

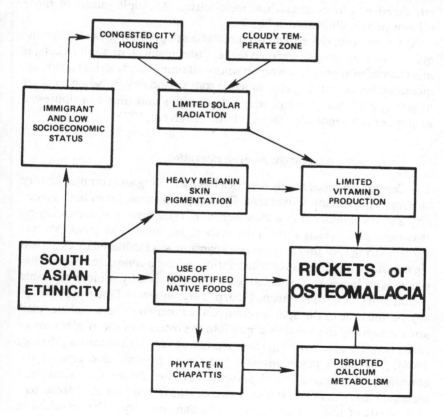

Fig. 7.5. A diagram indicating the conditions leading to rickets and osteomalacia in South Asian migrants to the United Kingdom.

of food products such as milk (Roe, 1985, p. 33). Tropical Indian or Pakistani immigrant children and pregnant women were at an adaptive disadvantage in United Kingdom cities for a number of reasons. First, at the genetic level, the dark melanin skin pigmentation, useful in the tropics as a shield against the damaging effects of ultraviolet radiation, now served to limit the synthesis of vitamin D in skin by normal actinic conversion of provitamin D. The reduced biological ability of South Asians to synthesize vitamin D was even worsened by their congested urban residences and new-immigrant ghettos where little sunlight and actinic ultraviolet radiation was found. As many immigrants to Western cities have found, only the poorest housing is available. Finally, an attempt to maintain a traditional South Asian diet cut off the immigrant from the vitamin-D-fortified Western foods, and by an unusual coincidence, the staple food, chapattis, contained a substance that further interfered with normal calcium metabolism. A simple model of these relationships is illustrated in Fig. 7.5.

As a final note, rickets and osteomalacia are probably selected against quite effectively, in a Darwinian sense. Osteomalacia, or adult rickets, is most common in pregnant women whose vitamin D and calcium needs are increased because of this reproductive status. And rickets, when it occurs in young girls, may lead to pelvic deformities and thus contributes to higher perinatal mortality during childbirth (Neer, 1975).

Integrated research: Andean migration

The Central Andes of South America has had a long and complex history of human population movement. During prehispanic times both voluntary and involuntary migrations were a normal means of exploiting an environment in which vertical diversity of resources was great (Murra, 1972). During Inca times, imperial expansion was facilitated by establishing *mitimaes*, or a process of colonization, where a conquered population was moved intact to a pacified area and replaced by a more tractable population (Steward & Faron, 1959, p. 127; Dobyns & Doughty, 1976, p. 48). At the time of the Spanish conquest, Francisco Pizarro and his small army moved up the western slopes into the interior of the Andes only to suffer considerable hardship from hypoxia at very high altitudes (Monge, 1948). The harsh environmental stresses of hypoxia and cold at the elevations of the *puna* and *altiplano* slowed, but did not halt, European incursion into the *sierra*. It was, however, the development of *Mestizos* – individuals of mixed European and Indian ancestry – that resulted in European 'genetic' occupation of the high-altitude zones (Dobyns &

Doughty, 1976, p. 123). During the Spanish colonial period, large numbers of families were forcibly relocated for mining labour at Huancavelica in Peru and Potosi in Bolivia (Escobar & Beall, 1982). Dobyns & Doughty (1976, p.103) estimate that three million men worked in the two mines over a period of 200 years, many being drawn from distances of up to 1000 km.

It was not until the nineteenth and twentieth centuries that population increases in the *sierra* zones and economic opportunities in the lowlands led to the modern pattern of downward migration. In Peru today, population increases in the eastern *montaña* and the coastal lowland zones are more than twice as great as increases in the central *sierra* zones (Wilkie & Perkal, 1984). This is largely the result of massive outmigration from the highlands to both the eastern and western lowlands. In Bolivia, eastward movement of people from the *altiplano* to middle and low elevations is comparable in scale (Escobar & Beall, 1982), and is a function of the government-supported pioneering settlement of the tropical lowlands.

In any context, human migration over considerable distances and from rural to urban settings produces a host of social, lifestyle and environmental changes that are very stressful. Nowhere are the changes greater than when individuals move from a high-altitude environment to the lowlands, or a low-altitude environment to the highlands. In these cases, stresses from elevation shifts require major physiological adjustments in addition to the more customary social and health adjustments of lateral migration. The widespread trend of downward migration has been exploited in several research programmes designed to explore problems of adaptation to high altitude and the effects of removal from this environment of native altitude-adapted individuals. Fig. 7.6 is a map for reference to the study locations.

Highlands and coast of southern Peru

The Peruvian research had as its original objective the study of the biocultural patterns of adaptation of Andean Quechua Indians to high altitude, hypoxic stress. Two large communities were studied intensively. The first study was conducted in the district of Nuñoa from 1964 to 1972 (Baker, 1969; Baker & Little, 1976). Nuñoa was a largely native Quechua-speaking district whose haciendas and native settlements range in elevation from about 4000 to 4600 m above sea level. The second study was conducted in the Tambo Valley in 1970 and 1971 (Baker, 1977a; Baker & Beall, 1982). The Tambo Valley was one of several irrigated

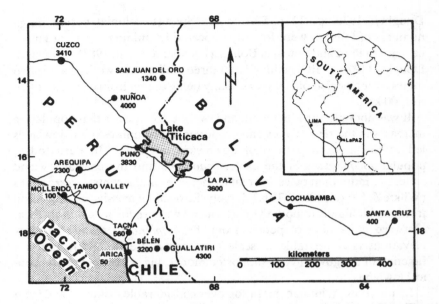

Fig. 7.6. Map of the central Andes at the borders of Peru, Bolivia and Chile. The numbers below the town and city names represent altitude above sea level in metres.

coastal valleys of southern Peru. Both the Nuñoa and and Tambo Valley areas were agrarian, but pastoral subsistence was also an important component of the Nuñoa economy. The coastal investigations in the Tambo Valley were derived ultimately from the Nuñoa work, and based on the research design developed from Harrison (1966) as described above.

Although it was not possible to follow the design exactly, three coastal subpopulations were compared with the high-altitude sedentary population of Nuñoa (high-altitude sedentes): (1) low-altitude-born native non-migrants (low-altitude sedentes), (2) low- and middle-altitude-born (<3000 m) migrants to the Tambo Valley (low-altitude migrants), and (3) high-altitude-born (>3000 m) migrants to the Tambo Valley (high-altitude migrants). Upward migrants were rare and not studied at this time. In this comparative design, it was assumed that both sedente groups were better adapted to their respective environments than were either of the two migrant groups. The group comparisons were expected to lead to the following: 1) high-altitude sedentes *vs.* high-altitude migrants would help to sort out effects of altitude acclimatization and genetic adaptation, particularly when length of residence at sea level was considered; 2) high-altitude migrants *vs.* low-altitude migrants would further define altitude effects, since both groups would have experienced residence change; 3) low-altitude migrants *vs.* low-altitude sedentes would define residence change effects. Some of the same measures of 'biological fitness' were evaluated in the Tambo Valley populations as in the earlier Nuñoa research. These measures were: fertility, mortality, health and disease, nutrition, child and adolescent growth, and work capacity.

Fertility

An interest in fertility effects of high altitude hypoxia was first stimulated by Monge (1948) when he documented the difficulties that the Spanish had in maintaining their numbers and reproducing at high altitude. Evidence for hypoxic effects on reproduction are based on animal studies and detailed literature review (Clegg, 1978) and correlations between human fertility and altitude of residence in Ecuador, Peru and Bolivia (James, 1966). Fertility surveys of women in Nuñoa (Hoff, 1968) and Tambo Valley (Abelson, Baker & Baker, 1974) suggest that high-altitude residence may limit fertility in native residents. High-altitude sedente women from Nuñoa had a total of 6.7 live births during their reproductive lives with a birth interval of about 5.0 years (for births 1–6). High-altitude migrant women from Tambo Valley had 8.46 live births with a birth interval of 2.7 years. The two low-altitude-born Tambo Valley groups had values that were intermediate. Since the sociocultural factors that influence human fertility are complex, these findings should be interpreted with some caution (Little & Baker, 1976). Nevertheless, the migration design *has* provided additional information toward the solution of this difficult problem.

Mortality

Respiratory disorders at high altitude carry a high risk of complications or death because they can severely limit the transfer of oxygen from the lungs to circulating blood. Respiratory disorders are also reported to be common at high altitude (Way, 1976a). Spector (1971) found that among more than 3000 deaths in the district of Nuñoa, about 60% were attributed to respiratory diseases, and Hoff's (1968) study of child mortality in the same area showed respiratory causes of death to be about 35% of total causes. The patterns of mortality of low-altitude natives (born at less than 1500 m above sea level), medium-altitude natives (born at between 1500 and 3000 m), and high-altitude natives (born above 3000 m), all residents at sea level in the Tambo Valley, are represented in Table 7.1. High-altitude migrants retain the high death rate from respiratory causes when

Table 7.1. *Mortality rates by cause-of-death[1] for three populations in Islay Province (region of Tambo Valley)*

Altitude of birthplace	n	Estimated mortality rates 1951–71				
		Accidental	Respiratory	Cardio-vascular	All other	Total
Low (sea-level–1500 m)	371	0.8	3.1	5.0	5.9	14.8
Medium (1500–3000 m)	392	1.7	4.6	5.6	10.7	22.6
High (greater than 3000 m)	237	2.6	6.9	3.9	10.2	23.6

[1]Age-specific deaths per year per 1000 individuals 45 years or older.
Data from Dutt & Baker, 1981.

at sea level, and also show the highest accidental death rate, but the lowest cause of death from cardiovascular disease. However, the cost of migration from high altitude may also be represented by the total crude death rate of 28.5/1000 which is slightly lower than the death rate for the district of Nuñoa (Baker & Dutt, 1972) but highest among the Tambo Valley residents.

Health and disease

When low-altitude sedente, low-altitude migrant and high-altitude migrant groups from Tambo Valley were compared, high-altitude male migrants had the greatest number of reported symptoms of illness and the greatest number of lost working days in a month (Dutt & Baker, 1978). The authors suggested that this pattern of illness reflected the stresses of migration from the highlands.

Blood pressure was surveyed in Nuñoa at high altitude and in the Tambo Valley at sea level. Of the migrant and sedente groups at sea level, the low-altitude sedentes and low-altitude migrants had the highest systolic and diastolic blood pressures whereas the middle-altitude and high-altitude migrants had the lowest values (Baker & Beall, 1982). Based on a Nuñoa blood pressure survey (Baker, 1969), it appears that the low values at altitude were retained by migrants to sea level. These results are counter to the customary pattern of increased blood pressure among migrant populations (Huizinga, 1972).

Nutrition

From Nuñoa studies, high-altitude dietary intake is highly variable in foods consumed, energy intake and specific nutrient intake (Gursky,

1969; Mazess & Baker, 1964; Picón-Reátegui, 1976). Tubers, grains and chenopods are the primary staples at high altitude on the *altiplano*. At sea level in the Tambo Valley, Riley (1979) found more diversity in foods consumed, probably because of the greater availability of both highland and lowland food products. Several conclusions were drawn about dietary status of high-altitude migrants (*N* = 52 households) with regard to low-altitude natives (*N* = 32 households) from the 1971 survey of Riley (1979). First, with the exception of low vitamin A (also low in high-altitude sedente diets; Picón-Reátegui, 1976), migrant diets seem adequate in calories and nutrients. However, newly migrant families (resident for less than two years) have slightly lower intakes than longer resident migrants. Migrants *do* change their food consumption habits rapidly toward local foods. Therefore, the lower food intakes in new migrants probably are indicative of economic instability during the initial several years in the new setting.

Child and adolescent growth
Infant, child and adolescent growth is a sensitive indicator of health and nutritional status. The effects of disease and malnutrition on growth have been well documented (Frisancho, 1980; Martorell, 1980), while effects of altitude on growth are less well known, but can be better understood by comparisons of highland sedentes and migrants. Haas (1976) developed a research design to control for residence, ethnicity and altitude in tests of psychomotor and physical development of infants up to 24 months of age. By an intricate series of comparisons of infants from Tambo Valley and the city of Tacna at sea level (first or second generation high-altitude migrants) and Nuñoa and the city of Puno (high-altitude sedentes), Haas (1976) concluded that motor development was most influenced by ethnicity and not altitude (Indian *vs*. Mestizo), while physical growth was clearly retarded at altitude (after controlling for residence and ethnicity). Among older children, Beall and her colleagues (1977) compared the physical growth of low-altitude sendente and high-altitude migrant children from Tambo with high-altitude sedente children from Nuñoa. They found that the growth of the Tambo low-altitude sedentes was advanced at all ages and achieved a greater adult size than the high-altitude sedente sample. The high-altitude migrant children were intermediate in their growth status. The two sedente samples are graphed for stature in Fig. 7.7 along with data from several other studies to illustrate the general pattern. The single exception to the slower growth at all ages of the high-altitude native children is in measures of chest size. Chest sizes of the two Tambo groups were equivalent to the Nuñoa group. Hence, relative to general size, the Nuñoa children's chest dimensions

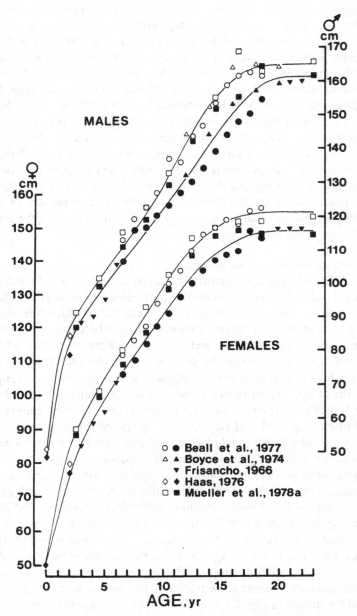

Fig. 7.7. Growth in stature of sedente high-altitude Indians and migrants to low altitude (open symbols are migrants to the lowlands; solid symbols are high-altitude sedentes).

were the largest of the three groups, probably as a function of a lifetime of high-altitude residence. In the only recorded study of growth of high-altitude migrant children to the eastern lowlands of Peru (San Juan del Oro), Hoff (1974) found that the high-altitude sedentes were smaller than the migrants in all but the chest dimensions. These results are compatible with the Tambo Valley studies.

Work capacity

Among visitors and recent migrants to high altitude the ability to perform physical work, as measured by the individual's aerobic capacity, is severely impaired. This is so, despite the fact that high-altitude natives can perform work or exercise as adequately in their home environment as can low-altitude natives at sea level (Buskirk, 1978). The research questions that were addressed to migrants were whether high-altitude migrants to sea level and low-altitude migrants to high altitude could achieve comparable degrees of work capacity as their sedente counterparts. Frisancho and his colleagues (1973) tested young males who had migrated to Cuzco, Peru, at different ages. He found that the maximal work capacity was correlated with the number of years residence at high altitude (positive) and the age at migration to high altitude (negative). Boys who had migrated to Cuzco between the ages of two and six years had maximal work capacities that were equivalent to native Quechua Indian residents. These results suggested that developmental acclimatization to hypoxia during childhood was necessary for full adjustment in work capacity. Way (1976b) studied maximal work capacity in high-altitude sedentes from Nuñoa and high-altitude migrants in the Tambo Valley. He found, as expected, that differences between high-altitude and low-altitude values for maximal work capcity were less for downward migrants (his study) than for upward migrants (other studies). His conclusions were that, although physical fitness variations could not be ruled out, the evidence suggested a genetic adaptation to hypoxia.

Highlands and lowlands of Bolivia and Chile

Two additional major research programmes have dealt with migrants and explored problems of high-altitude adaptation. In Bolivia, studies were conducted of maternal health and prenatal and postnatal growth at high altitude in La Paz and also among Quechua and Aymara downward migrants to the eastern savanna and forest lowland zone around Santa Cruz (Haas, 1980; Haas *et al.*, 1980, 1982). In Chile, work under a major research programme has focused on the genetics and health of native Aymara living along an altitudinal gradient from Arica at sea level to the highest elevations (4000–5000 m) of the Chilean *altiplano* (Cruz-Coke *et al.*, 1966; Mueller *et al.*, 1980; Schull & Rothhammer, 1977). Brief discussions of these research programmes follow.

Bolivian research

Haas' (1980) prenatal growth research design compared women of European and Indian (Quechua and Aymara) ancestry who were born at high or low altitude and then migrated to high-altitude (European) or low-altitude (Indian) during childhood or as an adult. Ethnicity was a primary variable in birthweight at high altitude and among migrants. Indian newborns were consistently heavier than non-Indian in La Paz despite the lower socioeconomic status of the Indian mothers (Haas *et al.*, 1980). Low-altitude birth weights of Indian babies were equivalent to or slightly greater than non-Indian babies. Haas (1980) hypothesized that it is the superior oxygen transport system of the Indian mothers that contributes to the higher birth weights and that this ability derives from a lifetime of exposure to hypoxia as well as selection for a genetically efficient reproductive system. Postnatal growth studies of La Paz and Santa Cruz infants up to one year of age demonstrated that, despite the prenatal advantage of high-altitude Indian mothers, Indian and Mestizo infants grow more slowly than ethnically comparable migrants at sea level (Haas *et al.*, 1982). Despite these more rapid growth patterns of downward Indian migrant infants, Foxman, Frerichs & Becht (1984) and Weil (1979) found high infant and young child mortality rates in eastern lowland migrant settlements.

Migrant boys of European ancestry were studied in La Paz during submaximal exercise tests (Haas *et al.*, 1983). Comparisons of 29 high-altitude-born and 26 low-altitude-born pre-adolescent boys indicated that the high-altitude-born were better adapted to the hypoxic conditions at 3600 m since they were able to perform an equivalent amount of work at less oxygen cost than the low-altitude-born boys. Greksa & Haas (1982) observed fundamentally the same pattern with a maximal exercise test.

Chilean research

Chilean investigations were conducted in the Department of Arica over an altitudinal gradient where large samples of subjects were examined at two coastal sites, five middle-altitude villages (2500–3500 m), and four *altiplano* hamlets (>4000 m). Much of the research dealt with permanent residents at each of the altitudinal levels and confirmed previous findings on growth of children (Mueller *et al.*, 1978a, 1980) and measures of lung function (Mueller *et al.*, 1978b, 1978c). Children and adult migrants were inspected in one study of chest size and lung function (Mueller *et al.*, 1979). In this study of 248 upward migrants and 129 downward migrants, the former group showed increased chest depth, forced vital capacity and one-minute forced expiratory volume, whereas the latter group showed no relationships when compared with sedentes at different elevations.

Synoptic findings

Based on investigations of migrants and sedentes at high and low altitudes, the following observations can be made.

Fertility may be impaired in high-altitude sedentes, since downward migrants to sea level experience reduced birth interval and increased completed fertility. That fertility is adequate at high altitude to maintain population numbers may result from the superior oxygen transport system of high-altitude Indian mothers.

Mortality is high in downward migrants in contrast to low-altitude sedentes, but cause of death is largely the result of accidents and respiratory diseases in adults. Low rates of hypertension and low risk of cardiovascular disease in older downward migrants from high altitude suggests that high-altitude residence in childhood and adolescence may reduce the overall risk of cardiovascular disease in later life. Infant and early childhood mortality is elevated in some high-altitude migrant groups.

As in the case of any population that moves, the new migrant is likely to suffer from economic problems that may be translated into a poorer dietary intake when compared with sedentes. This holds for high-altitude migrants.

Low-altitude sedentes are larger than high-altitude sedentes except for chest dimensions in which high-altitude natives are greater either absolutely or relative to stature. High-altitude migrants are intermediate between the two sedente populations.

Exercise capacity in low-altitude migrants to high altitude is usually impaired. It is suggested that the more complete adaptation in maximal work capacity requires a chronic exposure to altitude during childhood – that is, a pattern of developmental adaptation to high-altitude hypoxia. High-altitude migrants to sea level often perform better in tests of exercise capacity than low-altitude sedentes.

Integrated research: Pacific migration

The first truly long-distance migrants in the world were the ancient Pacific islanders who sailed from somewhere in Melanesia to the islands of Tonga and Samoa as early as 3500 years ago (Bellwood, 1978). These remarkable population movements, which continued up to several hundred years ago, resulted in the discovery of, and at least brief habitation of, every livable island in the Pacific. There were probably a variety of evolutionary processes acting on these migrant populations during three millennia of voyages and island settlement (Baker, 1984a). Since most exploration and pioneer settlement groups were small in number, the *founder effect* and *genetic drift* must have operated. Random processes associated with numerically small populations must also have been enhanced by accidents during and shortly after voyages and by the devastating effects of hurricanes on small islands and atolls. *Natural selection* was suggested as a process to have operated to favour large body size, obesity, and, perhaps, an efficient means of energy storage against the selective pressures of undernutrition or starvation during long voyages and the cold stress of night-time sailing (Baker, 1984a). Disease, as a selective pressure, played a major role in Pacific island depopulation during the period of early European contacts in the late eighteenth and nineteenth centuries. All of these and other factors contributed to the considerable variation and adaptation of Pacific island populations.

The pioneering migration of past Pacific islanders is quite different from the kinds of migration being experienced by contemporary Pacific natives. Today, Pacific populations have been relocated because of atomic testing (Bikini Atoll), phosphate exploration (Ocean Island), and a combination of limited resources and overpopulation (Tikopia, Gilbert Islands) (Lieber, 1977). Other Pacific populations are migrating voluntarily to larger, Westernized towns and major population centres of the Pacific because of population pressures, rising expectations associated with modernization, and perceived economic opportunities.

Studies of the effects of migration on human health and adaptability of Pacific natives were initiated in the late 1960s as a part of the New Zealand contribution to the International Biological Programme (Collins & Weiner, 1977, pp. 218–23). This research, the Tokelau Island Migrant Study, was designed as a long-term epidemiological exercise to investigate the development of new patterns of morbidity associated with Westernization in Tokelauan migrants to New Zealand (Prior, 1971, 1981; Prior *et al.*, 1974, 1977). A parallel programme of longitudinal research was initiated in the mid-1970s to study migrants from Samoa to Hawaii with regard to the effects of rapid environmental change on the biology, behaviour and health of the people (Baker, 1977b, 1984a). Results of these two major projects and additional work with Pacific migrants are described in the following paragraphs. Fig. 7.8 provides locations of the Pacific islands that have been sites of immigration and emigration.

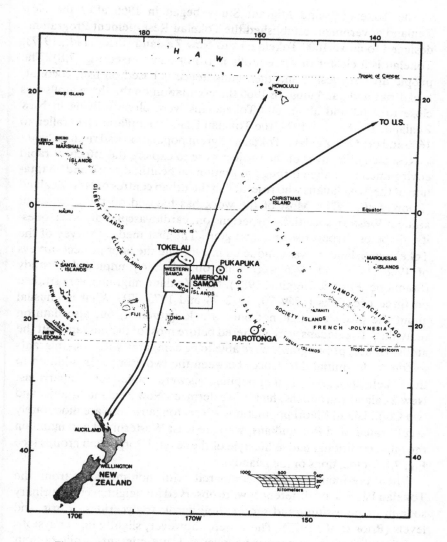

Fig. 7.8. Map of the Pacific Ocean Islands on which migrant population studies have been conducted.

The Tokelau Island Migrant Study

The Tokelau Island Migrant Study began in 1966 after the New Zealand government established the Tokelau Resettlement Programme designed to move 1000 Tokelauans to New Zealand (Prior *et al.*, 1977). Tokelau is a cluster of three atolls north of Samoa (see Fig. 7.8). The people are Polynesian and their subsistence is based on fish, coconut, breadfruit and pandanus. In 1966 the population on the three atolls was close to 2000 and about 500 Tokelauans were already living in New Zealand. By the year 1974, the Tokelau Islands population had fallen to 1600 and the New Zealand Tokelau migrant population had risen to 2000.

Research objectives of the project were to explore the effects of rapid environmental change through migration on health, disease and adaptation of the Tokelauans who resettled in the urban centres of New Zealand (Prior *et al.*, 1977). Most of the work has focused on the diseases of affluent Western societies – hypertension, cardiovascular diseases, obesity, diabetes, hyperurisaemia and gout. The first medical survey of the Tokelauan Islanders was conducted in 1963 and the later project surveys of 1968, 1971 and 1976 were built on this early unpublished study (Stanhope, Prior & Fleming, 1981). New Zealand migrants were studied on three occasions (1969–70, 1972–74 and 1975–77). What is unusual about the project is that not only is it longitudinal, but also data on individual Tokelauans were gathered before most of the migrants left the atolls. Hence, pre-migrants were able to be compared with non-migrants to control for initial differences between the two groups. In addition to the Tokelauans, several other populations were studied for comparisons: New Zealand Europeans, highly Westernized New Zealand Maoris, and two Cook Island Maori populations – Rarotongans, who are moderately acculturated, and Pukapukans, who are least Westernized and maintain the most traditional native lifestyle of the several Polynesian groups (see Fig. 7.8 for locations of the Islands).

When pre-migrants were compared with non-migrants from the Tokelau Islands, no differences were observed for height, weight, urinary sodium or potassium, and serum cholesterol, triglycerides or uric acid levels (Prior *et al.*, 1977). There were, however, slightly higher systolic and diastolic blood pressures in younger male pre-migrants, while 24-hour urinary sodium values were generally lower in pre-migrants when compared with non-migrants. The higher blood pressures in pre-migrants, as noted by Prior and his colleagues (1977), might have reflected a more acculturated or Westernized status or fundamental differences in personality between the two groups.

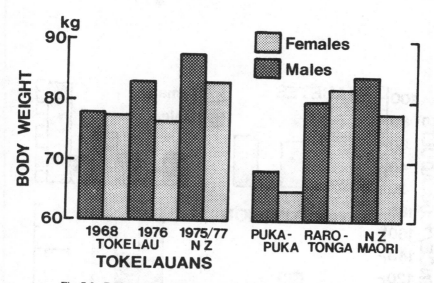

Fig. 7.9. Body weights of Tokelauan migrants and sedentes compared with three other Polynesian populations. (Data from Prior, 1981; Prior *et al.*, 1981; Finau *et al.*, 1982.)

Some of the baseline data for the migrant studies was provided by comparing diets of Tokelauans and Pukapukans on their home atolls (Prior *et al.*, 1981). Both diets were traditional Polynesian with a high proportion of energy being provided by coconuts. Coconuts provided 34% of Pukapukan dietary energy and 63% of Tokelau energy, leading to a higher intake of saturated fat by Tokelauans and, probably, their higher serum cholesterol levels as well. Despite the high intake of dietary fat (about 35% of calories in Pukapukans and 53% of calories in Tokelauans), cardiovascular disease was in low prevalence in both populations. Body weight differences were distinct between the two island groups.

Fig. 7.9 illustrates body weight differences between Tokelauans and Pukapukans along with Tokelauan migrants, New Zealand Maori and Rarotongans. The Westernized New Zealand Maori, Rarotongans and Tokelauan migrants to New Zealand displayed the highest body weights, Tokelau Islanders were intermediate and Pukapukans were lowest in weight. With the exception of the Pukapuka sample, the Polynesian groups showed very high body weights associated with obesity and high skinfold values (Finau, Prior & Evans, 1982). Prior (1981) provided more detailed, age-standardized data that demonstrated body weight increases

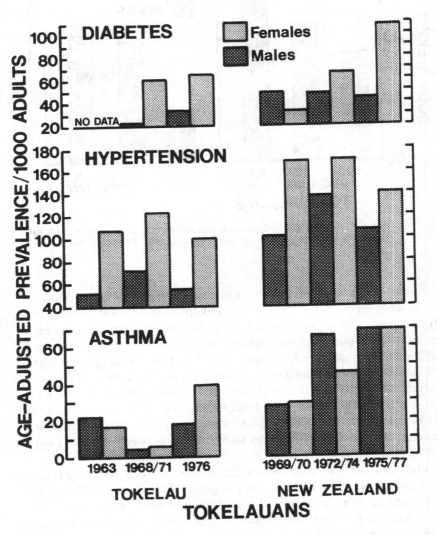

Fig. 7.10. Prevalence of three disorders among Tokelauan sedentes and migrants to New Zealand. (Data from Stanhope *et al.*, 1981.)

in Tokelauan Islanders between the years 1963 and 1976, but even greater body weight gains in the Tokelau migrants to New Zealand, including children (Ramirez & Mueller, 1980).

High saturated fat intake and obesity predispose Polynesians to a variety of chronic diseases including cardiovascular disorders, hypertension and diabetes (Harvey & Prior, 1972; Stanhope *et al.*, 1981; Zimmet & Björntorp, 1979). When Islanders migrate to Western cities or become Westernized through *in situ* acculturation, these risk factors increase, as does general morbidity. Three disorders representing some of these increases in morbidity among Tokelau migrants are presented in Fig. 7.10. Diabetes had increased only slightly in Tokelauan Islanders and male migrant Tokelauans, whereas female migrants had shown marked increases in prevalence in New Zealand. Asthma had also increased in prevalence among the migrants, but Stanhope and others (1981) suggested that the high prevalence was a recurrence of childhood asthma, perhaps as the result of the novel and stressful environment. Hypertension (systolic blood pressure > 160 torr or diastolic pressure > 95 torr) prevalence rates in Tokelauan migrants were substantially elevated over rates observed in the Tokelau Island population.

One of the most consistent patterns of occult morbidity in migrant populations is increased blood pressure and associated hypertension. Comparisons of several Polynesian groups by age and sex are made for blood pressure in Fig. 7.11. The least acculturated Pukapukans had

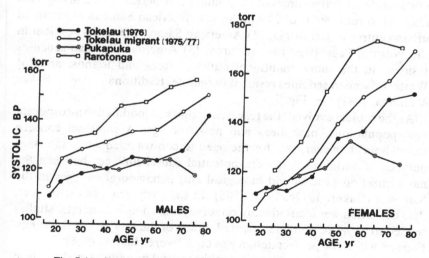

Fig. 7.11. Systolic blood pressures of Tokelau sedentes and migrants, Pukapukans and Rarotongans. (After Prior, 1981; Finau *et al.*, 1982.)

lowest systolic blood pressures and least age changes. Tokelauan Islanders had values close to those of the Pukapukan Islanders, but blood pressure increases with age were present, particularly in women. Rarotongan Islanders' blood pressures were higher than all of the other populations and Tokelauan migrants to New Zealand were intermediate. Although the greater body weight of the Tokelauan migrants was statistically correlated with higher blood pressures in that group, this relationship explained only a portion of the variance in the differences between the migrants and the non-migrants (Joseph *et al.*, 1983). Problems of acculturation and blood pressure increases will be discussed further in the context of Samoan migrations.

The Samoan Migrant Study

The Samoan Migrant Study was initiated in 1975 with the general objective to study the consequences of modernization through Samoan migration and '. . . to examine how Samoan utilization of the natural and social environment related to their current socioeconomic, behavioral, and biological characteristics'. (Baker & Hanna, 1981; Baker, 1987). In the decade between 1975 and 1985 several subpopulations of Samoan migrants and sedentes were investigated in order to sort out those environmental parameters that influenced biological fitness and adaptability in Samoan populations. These were: (1) American Samoan migrants resident in urban areas of Honolulu, (2) American Samoan migrants resident in rural areas of Honolulu, (3) American Samoan migrants to urban centres in California, (4) American Samoan sedentes resident in the Western Pago Pago harbour area, (5) American Samoan sedentes resident in the more traditional outlying areas and islands, and (6) Western Samoan sedentes resident in remote, traditional villages (Baker & Hanna, 1981) (see Fig. 7.8).

As the project evolved, the research design was modified to incorporate new populations, new ideas and new findings. A principal concern throughout, however, was for the need to control adequately the large number of variables that were potential causal linkages between the modernization process and biological and behavioural changes in the Samoans (Baker, 1977c, 1981, 1985). In the early years of the project, data collection was focused more on survey and comparison: later studies tended to be more problem oriented. Some of the fundamental assumptions on which the project design was based were (Baker, 1981):

(1) Changes in the environment, either social or natural, will produce stresses that will adversely affect a population's health.

(2) The more dramatic or the more rapid the environmental change, the greater the adverse effects.

(3) Since developmental adjustments to environmental change play an important role in adaptation, then younger individuals are likely to suffer from less stress or health impairment than older individuals.

(4) The adverse effects of migration should be most intense during the early period following migration or at least during the first generation migrants.

In the section which follows, only those aspects bearing on obesity and cardiovascular disease are examined to illustrate the problems of distinguishing between the effects of modernization and migration.

One major attribute of Samoan biological status is obesity (Baker, 1985; Bindon & Baker, 1985; Pawson & Janes, 1981). Fig. 7.12 shows the trend in body weight from the least modernized Western Samoans to successively more modernized American Hawaiian and Californian Samoans. The trend of increasing body weight with increasing modernization held for both sexes. Yet when triceps skinfold was used as a measure of adiposity or changing obesity, then there was only a slight effect of migration on obesity in men and no change in women. The only distinct difference in skinfold thickness was between Western Samoan rural natives and the other Samoan groups, all of whom had experienced some degree of modernization. Hence, although it is clear that changes in lifestyle were linked to obesity in Samoans, migration alone did not produce an increase in obesity prevalence.

Nevertheless, obesity was in high prevalence among all but the most traditional Samoan populations and this placed them at greater risk of morbidity from cardiovascular diseases and diabetes. Obesity as a risk factor was exacerbated by the reported low levels of physical fitness of Samoans. American Samoan men had aerobic capacities that were considerably lower than United States cohorts (Baker & Hanna, 1981). Low aerobic capacities were also associated with elevated body fatness (Greksa & Baker, 1982), and body fatness was positively and highly correlated with systolic and diastolic blood pressures (Hanna & Baker, 1979; McGarvey & Baker, 1979). When systolic blood pressures were compared for two migrant and two sedente Samoan populations (Fig. 7.13), the migrant populations showed slightly elevated blood pressures, but only when the more traditional American Samoans constituted this sedente sample. Also, it was only the Hawaiian migrants from the more traditional areas of Samoa who had higher blood pressure than American Samoan sedentes (McGarvey & Baker, 1979). What appears to be the case is that those sedentes who had become somewhat modernized before

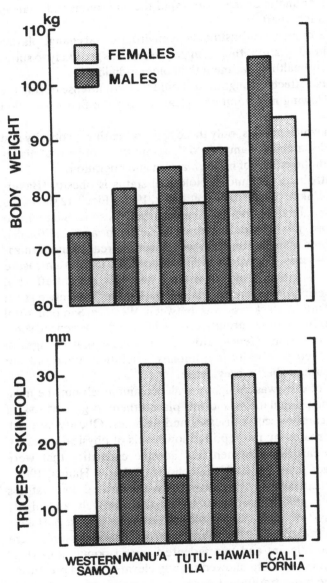

Fig. 7.12. Body weight and triceps skinfolds of Western Samoans, American Samoans (Islands of Manu'a and Tutuila), and American Samoan migrants to Hawaii and California, USA. (After Bindon & Baker, 1985; Pawson & Janes, 1981.)

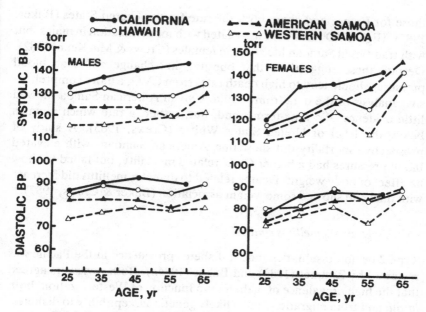

Fig. 7.13. Systolic and diastolic blood pressures of Western Samoan and tradi-tional American Samoan sedentes and American Samoan migrants to Hawaii and California. (After Baker & Hanna, 1981; Pawson & Janes. 1981.)

migrating had already been exposed to conditions predisposing them to blood pressure elevation. Thus, only by having very detailed information on degree of Westernization in the pre-migrant and subsequent exposure to Westernization in the migrant can some of these relationships be specified.

Cardiovascular disease (CVD), coronary risk factors and diabetes have been investigated in sedentary and migrant Samoans. Hornick & Hanna (1982) found that modernized Samoans had serum lipid values that placed them in slightly higher risk category for CVD than more traditional Samoans. However, a final analysis of all the studies on blood lipids showed that all Samoan groups were at relatively low CVD risk (Pelletier & Hornick, 1987). Pawson & Janes (1982) in their studies of California Samoans found a high mortality from CVD in adults over 50 years of age. A more comprehensive study of Samoan CVD and diabetes was con-ducted on American Samoan sedentes (Crews & MacKeen, 1982). It was found that American Samoa CVD death rates were on the rise from 1963 to 1974, and that age-adjusted CVD death rates were slightly greater than those for France and about 66% the rate for the United States (Baker, 1985). High CVD rates were associated with modernization in males, but with traditional Samoan life styles in females (Crews & MacKeen, 1982). Despite these suggestions that obesity, diet change and high blood pressures should lead to high death rates from CVD, a more comprehensive analysis of the data from 1950 to 1981 in American Samoa showed little evidence of a long-term trend, but, rather, a rate which remains below the level of United States Whites (Crews, 1985). A study of prospective mortality did show that American Samoans with elevated blood pressures had a higher CVD-related mortality, but failed to show an effect of body weight. Deaths related to diabetes mellitus did increase with time and showed some weight association (Baker & Crews, 1987).

Diabetes mellitus studies

Type 2 or non-insulin-dependent diabetes prevalence in the Pacific was reviewed by Zimmet (1982) and Baker (1984a). Both reviewers agreed that the high prevalence of diabetes is a function of Westernization, both *in situ* and from migration, and a likely genetic susceptibility to diabetes. The etiology of the disease in the Pacific, however, is far from clear. Although obesity is certainly implicated in the disease (Foster & Burton, 1985), diabetes prevalence is low in non-obese populations (Pukapuka; Prior *et al.*, 1966) and in populations where there is a moderate degree of obesity (Tuvalu; Taylor & Zimmet, 1981). Zimmet and others (1977) have uncovered what is perhaps the highest prevalence rate in the world: 34.4% diabetes in individuals over 15 years of age on Nauru in the Ellice Islands. Nauru is also noted as having one of the highest per capita incomes in the world because of the rich phosphate deposits (guano) on the island. These high prevalence values for diabetes in the Pacific are similar in character to the high rates among Amerindians in North and South America. Weiss, Ferrell & Hanis (1984) have referred to this pattern of Westernization, obesity and type 2 diabetes as a New World Syndrome. Perhaps Pacific populations should be included.

Synoptic findings

Based on the Pacific (largely Polynesian) studies of migrants to western centres, a number of tentative generalizations can be presented. South Pacific Islanders' traditional diet was high in saturated fat (from coconuts) yet body weights in adults were maintained at moderate levels.

Post World War II modernization – including a lifestyle modified by imported food, ideas and migration to Western areas – is associated with a constellation of disorders including chronic obesity, hypertension, hyperurisaemia, asthma, serum coronary risk factors and diabetes mellitus. Some of these disorders – obesity and diabetes – have far exceeded prevalence values among Western urban residents, and are thought to have a genetic basis.

There are indications from these studies that the declines in health and biological fitness of South Pacific migrants are not a direct effect of migration *per se,* but depend on the degree of social and environmental change that the migrant experiences. If, for example, the pre-migrant has already moved to a more modernized lifestyle, then the impact of migration is likely to be less intense.

Discussion

Study of the biological and adaptive attributes of migration is an important endeavour for at least two reasons. First, at the level of science and basic knowledge, migrant peoples serve as ideal natural experiments in which the effects of environmental change on human populations can be investigated within the theoretical framework of adaptation (Baker, 1966, 1977*c*; Kasl & Berkman, 1983). Secondly, there are real risks to human health and biological fitness associated with the process of migration, and these risks should be known and understood (Hull, 1979; Thomas, 1979). This latter, applied, basis for studies of migrants is especially important now, since present migration rates are greater than at any other time in the past (Davis, 1974).

However, there are a number of methodological problems that must be solved in order to strengthen future studies of migrants – both basic and applied studies. For example, the number of variables that have an impact on health or biological fitness of members of a population is large, and difficult to control (Boyden, 1972; Harrison, 1973; Baker, 1981). In the case of an acute disorder – rickets – in South Asian migrants, there were a limited number of variables to consider. Accordingly, an explanation for the disorder was provided and action was taken to treat the health problem (Donovan, 1984). On the other hand, elevated blood pressure, as a chronic disorder, is probably slow to develop, may involve a genetic predisposition, and is likely to result from behavioural variations associated with a modern lifestyle (Ward, 1983). The ability to partition out the variance in blood pressure from some of the many causally linked variables by use of migrants and non-migrants has not been entirely successful. This is partly due to the difficulty in quantifying the long-term sociocultural and lifestyle attributes that contribute to the physiological effect of high blood pressure. Studies of non-Western sedente populations *have* served the important function of identifying healthy standards for low or normotensive blood pressures (Huizinga, 1972; Mugambi & Little, 1983).

Another methodological problem involves the sorting out of changes produced by migration from those produced by modernization (Baker, 1981). The majority of migrants today have moved from a less modernized setting to a more modernized, and probably urbanized, environment (Hull, 1979). One of the real values of migration models of investigation is that the process of migration often produces very abrupt changes requiring rapid adaptation to the new social and natural environment. An important consideration in assessing the abruptness of the change is how much prior modernization had occurred before the people moved.

Finally, a major difficulty of migration studies is in generalizing and developing theory since each migration event may be viewed as unique (Kasl & Berkman, 1983). Andean Indian migrants from rural or urban highland settings may move to lowland *barriadas* or shantytowns at the outskirts of major cities, rural coastal farms or towns, or rural tropical forest homestead farms. This pattern alone produces two sedente populations, six migrant populations and three host populations and makes it incredibly difficult to implement a research design. What is needed is an integrated body of theory that combines current knowledge on culture change, decision-making, demography of migration and biological adaptation. Some beginnings have been made (Baker, 1981; Kasl & Berkman, 1983), but much remains to be done.

Acknowledgment. We wish to thank Ms Peg Roe for her assistance in manuscript preparation.

References

Abelson, A. E., Baker, T. S. & Baker, P. T. (1974). Altitude, migration and fertility in the Andes. *Social Biology*, **21**, 12–27.

Baker, P. T. (1966). Human biological variation as an adaptive response to the environment. *Eugenics Quarterly*, **13(2)**, 81–91.

Baker, P. T. (1969). Human adaptation to high altitude. *Science*, **163**, 1149–56.

Baker, P. T. (1976). Research strategies in population biology and environmental stress. In *The Measures of Man: Methodologies in Biological Anthropology*, ed. E. Giles & J. S. Friedlaender, pp. 230–59. Cambridge, Mass.: Peabody Museum Press.

Baker, P. T. (1977*a*). Un estudio de los aspectos biológicos y sociales de la migración Andina. *Archivos de Biología Andina*, 7, 63–82.

Baker, P. T. (1977*b*). Environment and migration on the small islands of the South Pacific. In *Human Population Problems in the Biosphere: Some Research Strategies and Designs*, ed. P. T. Baker, pp. 53–64. MAB Technical Notes 3. Paris: UNESCO.

Baker, P. T. (1977*c*). Problems and strategies. In *Human Population Problems in the Biosphere: Some Research Strategies and Designs*, ed. P. T. Baker, pp. 11–32. MAB Technical Notes 3. Paris: UNESCO.

Baker, P. T. (1981). Migration and human adaptation. In *Migration, Adaptation and Health in the Pacific*, ed. C. Fleming & I. Prior, pp. 3–13. Wellington, New Zealand: Wellington Hospital Epidemiology Unit.

Baker, P. T. (1984*a*). Migrations, genetics and the degenerative diseases of South Pacific Islanders. In *Migration and Mobility: Biosocial Aspects of Human Movement*, ed. A. J. Boyce, pp. 209–39. Symposia of the Society for the Study of Human Biology, **23**. London: Taylor and Francis.

Baker, P. T. (1984*b*). The adaptive limits of human populations. *Man*, **19**, 1–14.

Baker, P. T. (1985). Modernization, migration and health: a methodological puzzle with examples from the Samoans. S. S. Sarker Memorial Lecture.

Baker, P. T. (1987). Rationale and research design. In *The Changing Samoans: Health and Behavior in Transition*, ed. P. T. Baker, J. M. Hanna & T. S. Baker. New York: Oxford University Press.

Baker, P. T. & Baker, T. S. (1984). Tropical land management through migration: the consequences for Andeans and South Pacific Islanders. In *Ecology in Practice*, vol. I, ed. F. di Castri, F. W. G. Baker & M. Hadley, pp. 180–201. Dublin: Tycooly International Publishing.

Baker, P. T. & Beall, C. M. (1982). The biology and health of Andean migrants: a case study in south coastal Peru. *Mountain Research and Development*, **2(1)**, 81–95.

Baker, P. T. & Crews, D. E. (1987). Mortality patterns and some biological predictors. In *The Changing Samoans: Health and Behavior in Transition*, ed. P. T. Baker, J. M. Hanna & T. S. Baker. New York: Oxford University Press.

Baker, P. T. & Dutt, J. S. (1972). Demographic variables as measures of biological adaptation: a case study of high altitude human populations. In *The Structure of Human Populations*, ed. G. A. Harrison & A. J. Boyce, pp. 352–78. Oxford: Clarendon Press.

Baker, P. T. & Hanna, J. M. (1981). Modernization and the biological fitness of Samoans: a progress report on a research program. In *Migration, Adaptation and Health in the Pacific*, ed. C. Fleming & I. Prior, pp. 14–26. Wellington, New Zealand: Wellington Hospital Epidemiology Unit.

Baker, P. T. & Little, M. A. (eds) (1976). *Man in the Andes: A Multidisciplinary Study of High-Altitude Quechua*. Stroudsburg, Penn.: Dowden, Hutchinson and Ross.

Beall, C. M. (1982). An historical perspective on studies of human growth and development in extreme environments. In *A History of American Physical*

Anthropology 1930–1980, ed. F. Spencer, pp. 447–65. New York: Academic Press.

Beall, C. M., Baker, P. T., Baker, T. S. & Haas, J. D. (1977). The effects of high altitude on adolescent growth in southern Peruvian Amerindians. *Human Biology*, **49(2)**, 109–24.

Beiser, M., Collomb, H., Ravel, J. L. & Nafziger, C. J. (1976). Systemic blood pressure studies among the Serer of Senegal. *Journal of Chronic Diseases*, **29**, 371.

Bellwood, P. (1978). *The Polynesians: Prehistory of An Island People*. London: Thames and Hudson.

Bindon, J. R. & Baker, P. T. (1985). Modernization and obesity among Samoan adults. *Annals of Human Biology*, **12(1)**, 67–76.

Boas, F. (1912). *Changes in the Bodily Form of Descendants of Migrants*. New York: Columbia University Press.

Boyce, A. J., Haight, J. S. J., Rimmer, D. B. & Harrison, G. A. (1974). Respiratory function in Peruvian Quechua Indians. *Annals of Human Biology*, **1**, 137–48.

Boyden, S. (1972). Biological determinants of optimal health. In *Human Biology of Environmental Change*, ed. D. J. M. Vorster, pp. 3–11. London: International Biological Programme.

Briggs, A. (1983). The environment of the city. In *How Humans Adapt: A Biocultural Odyssey*, ed. D. J. Ortner, pp. 371–94. Washington, DC: Smithsonian Institution Press.

Brown, D. E. (1981). General stress in anthropological fieldwork. *American Anthropologist*, **83**, 74–92.

Brown, D. E. (1982). Physiological stress and culture change in a group of Filipino-Americans: a preliminary investigation. *Annals of Human Biology*, **9(6)**, 553–63.

Buskirk, E. R. (1978). Work capacity of high-altitude natives. In *The Biology of High-Altitude Peoples*, ed. P. T. Baker, pp. 173–87. Cambridge: Cambridge University Press.

Carruthers, M. (1976). Biochemical responses to environmental stress. In *Man in Urban Environments*, ed. G. A. Harrison & J. B. Gibson, pp. 247–73. Oxford: Oxford University Press.

Chamberlain, M. L. & Hosking, D. J. (1971). Nutritional osteomalacia in immigrants. *Lancet*, **2**, 603–4.

Childe, V. G. (1952). *Man Makes Himself*. New York: New American Library.

Clark, F., Simpson, W. & Young, J. R. (1972). Osteomalacia in immigrants from the Indian subcontinent in Newcastle upon Tyne. *Proceedings, Royal Society of Medicine*, **65**, 478–80.

Clark, J. I. (1984). Mobility, location and society. In *Migration and Mobility: Biosocial Aspects of Human Movement*, ed. A. J. Boyce, pp. 355–70. London: Taylor and Francis.

Clegg, E. J. (1978). Fertility and early growth. In *The Biology of High-Altitude Peoples*, ed. P. T. Baker, pp. 65–115. Cambridge: Cambridge University Press.

Collins, K. J. & Weiner, J. S. (1977). *Human Adaptability: A History and Compendium of Research in the International Biological Programme*. London: Taylor and Francis.

Crews, D. E. (1985). *Mortality, Survivorship and Longevity in American Samoa,*

1950–1981. PhD Dissertation in Anthropology. University Park, Pa. Pennsylvania State University.

Crews, D. E. & MacKeen, P. C. (1982). Mortality related to cardiovascular disease and diabetes mellitus in a modernizing population. *Social Science and Medicine,* **16,** 175–81.

Crosby, A. W., Jr (1973). *The Columbian Exchange: Biological and Cultural Consequences of 1492.* Westport, Conn.: Greenwood Press.

Cruz-Coke, R., Cristoffanini, A. P., Aspillaga, M. & Biancani, F. (1966). Evolutionary forces in human populations in an environmental gradient in Arica, Chile. *Human Biology,* **38,** 421–38.

Davis, K. (1974). The migrations of human populations. *Scientific American,* **231(3),** 93–105.

Dobyns, H. F. & Doughty, P. L. (1976). *Peru: A Cultural History.* New York: Oxford University Press.

Donovan, J. L. (1984). Ethnicity and health: a research review. *Social Science and Medicine,* **19,** 663–70.

Dressler, W. W. (1982). *Hypertension and Culture Change: Acculturation and Disease in the West Indies.* South Salem, New York: Redgrave Publishing.

Dutt, J. S. & Baker, P. T. (1978). Environment, migration and health in southern Peru. *Social Science and Medicine,* **12,** 29–38.

Dutt, J. S. & Baker, P. T. (1981). An analysis of adult mortality causes among migrants from altitude and sedentes in coastal southern Peru. In *Health in the Andes,* ed. J. W. Bastien & J. Donahue, pp. 92–102. Special Publication **12.** Washington, DC: American Anthropological Association.

Eagan, C. J. (1967). The responses to finger cooling of Alaskan Eskimos after nine months of urban life in a temperate climate. *Biometeorology,* **2(2),** 822–30.

Escobar, M. G. & Beall, C. M. (1982). Contemporary patterns of migration in the Central Andes. *Mountain Research and Development,* **2(1),** 63–80.

Eveleth, P. B. (1966). The effects of climate on growth. *Annals of the New York Academy of Sciences,* **134,** 750–9.

Eveleth, P. B. & Tanner, J. M. (1976). *Worldwide Variation in Human Growth.* Cambridge: Cambridge University Press.

Finau, S. A., Prior, I. A. M. & Evans, J. G. (1982). Aging in the South Pacific: physical changes with urbanization. *Social Science and Medicine,* **16,** 1539–49.

Ford, J. A., Colhoun, E. M., McIntosh, W. B. & Dunnigan, M. G. (1972). Biochemical response of late rickets and osteomalacia to a chupatty-free diet. *British Medical Journal,* **3,** 446–7.

Foster, W. R. & Burton, B. T. (eds) (1985). Health Implications of Obesity: National Institute of Health Consensus Development Conference. *Annals of Internal Medicine,* **103(6–Part 2),** 977–1077.

Foxman, B., Frerichs, R. R. & Becht, J. N. (1984). Health status of migrants. *Human Biology,* **56,** 129–41.

Frisancho, A. R. (1966). *Human Growth in a High Altitude Peruvian Population.* MA Thesis in Anthropology. University Park, Pa.: Pennsylvania State University.

Frisancho, A. R. (1979). *Human Adaptation: A Functional Interpretation.* St Louis: C. V. Mosby.

Frisancho, A. R. (1980). Role of calorie and protein reserves on human growth during childhood and adolescence in a Mestizo Peruvian population. In *Social and Biological Predictors of Nutritional Status, Physical Growth, and Neurological Development*, ed. L. S. Greene & F. E. Johnston, pp. 49–58. New York: Academic Press.

Frisancho, A. R., Martinez, C., Velásquez, T., Sanchez, J. & Montoye, H. (1973). Influence of developmental adaptation on aerobic capacity of high altitude. *Journal of Applied Physiology*, 34, 176–80.

Ganong, W. F. (1981). *Review of medical Physiology*, 10th ed. Los Altos, Cal.: Lang Medical Publishers.

Gibson, G. S. & Gibbons, A. (eds) (1982). Hypertension Among Blacks: An Annotated Bibliography. *Hypertension*, II, 4(1), I-1–52.

Goel, K. M., Logan, R. W., Arneil, G. C., Sweet, E. M., Warren, J. M. & Shanks, R. A. (1976). Florid and subclinical rickets among immigrant children in Glasgow. *Lancet*, 1, 1141–5.

Goel, K. M., Campbell, S., Logan, R. W., Sweet, E. M., Attenburrow, A. & Arneil, G. C. (1981). Reduced prevalence of rickets in Asian children in Glasgow. *Lancet*, 1, 405.

Goldstein, M. S. (1943). *Demographic and Bodily Changes in Descendents of Mexican Immigrants*. Austin: Institute of Latin American Studies, University of Texas.

Graves, N. B. & Graves, T. D. (1974). Adaptive strategies in urban migration. *Annual Review of Anthropology*, 3, 117–51.

Greksa, L. P. & Baker, P. T. (1982). Aerobic capacity of modernizing Samoan men. *Human Biology*, 54(4), 777–88.

Greksa, L. P. & Haas, J. D. (1982). Physical growth and maximal work capacity in preadolescent boys at high altitude. *Human Biology*, 54, 677–95.

Greulich, W. W. (1957). A comparison of the physical growth and development of American-born and native Japanese children. *American Journal of Physical Anthropology*, 15, 489–515.

Gursky, M. J. (1969). *Dietary Survey of Three Peruvian Highland Communities*. MA Thesis in Anthropology. University Park, Pa.: Pennsylvania State University.

Haas, J. D. (1976). Prenatal and infant growth and development. In *Man in the Andes: A Multidisciplinary Study of High-Altitude Quechua*, ed. P. T. Baker & M. A. Little, pp. 161–79. Stroudsburg, Penn.: Dowden, Hutchinson and Ross.

Haas, J. D. (1980). Maternal adaptation and fetal growth at high altitude in Bolivia. In *Social and Biological Predictors of Nutritional Status, Physical Growth and Neurological Development*, ed. L. S. Greene & F. E. Johnston, pp. 257–90. New York: Academic Press.

Haas, J. D., Frongillo, E. A., Jr, Stepick, C. D., Beard, J. L. & Hurtado G., L. (1980). Altitude, ethnic and sex difference in birth weight and length in Bolivia. *Human Biology*, 52(3), 459–77.

Haas, J. D., Greksa, L. P., Leatherman, T. L., Spielvogel, H., Paredes, Fernandes, L., Moreno-Black, G. & Paz-Zamora, M. (1983). Submaximal work performance of native and migrant preadolescent boys at high altitude. *Human Biology*, 55, 517–27.

Haas, J. D., Moreno-Black, G., Frongillo, E. A., Jr, Pabon, A., Pareja L., G.,

Ybarnegaray, U. J. & Hurtado G., L. (1982). Altitude and infant growth in Bolivia: a longitudinal study. *American Journal of Physical Anthropology*, **59**, 251–62.

Hanna, J. M. & Baker, P. T. (1979). Biocultural correlates to the blood pressure of Samoan migrants in Hawaii. *Human Biology*, **51(4)**, 481–97.

Harrison, G. A. (1966). Human adaptability with reference to the IBP proposals for high altitude research. In *The Biology of Human Adaptability*, ed. P. T. Baker & J. S. Weiner, pp. 509–19. Oxford: Clarendon Press.

Harrison, G. A. (1973). The effects of modern living. *Journal of Biosocial Science*, **5**, 217–28.

Harrison, G. A. & Gibson, J. B. (eds). (1976). *Man in Urban Environments*. Oxford: Oxford University Press.

Harrison, G. A., Weiner, J. S., Tanner, J. M. & Barnicot, N. A. (1977). *Human Biology: An Introduction to Human Evolution, Variation, Growth, and Ecology*, 2nd edn. Oxford: Oxford University Press.

Harvey, H. P. B. & Prior, I. A. M. (1972). Cardiovascular epidemiology in New Zealand and the South Pacific. In *Human Biology of Environmental Change*, ed. D. J. M. Vorster, pp. 80–6. London: International Biological Programme.

Hiernaux, J. (1964). Weight/height relationship during growth in Africans and Europeans. *Human Biology*, **16**, 273–93.

Hornick, C. A. & Hanna, J. M. (1982). Indicators of coronary risk in a migrating Samoan population. *Medical Anthropology*, **6**, 71–9.

Hoff, C. J. (1968). Reproduction and viability in a highland Peruvian Indian population. *Occasional Papers in Anthropology* (University Park, Pa.: Pennsylvania State University), **1**, 85–161.

Hoff, C. J. (1974). Altitudinal variations in the physical growth and development of Peruvian Quechua. *Homo*, **24**, 87–99.

Hong, S. K. (1963). Comparison of diving and non-diving women of Korea. *Federation Proceedings*, **22**, 831–3.

Hori, S., Taguchi, S. & Horvath, S. M. (1975). Physiological responses to heat. In *Comparative Studies on Human Adaptability of Japanese, Caucasians and Japanese Americans, JIBP Synthesis, Human Adaptability*, vol. 1, ed. S. M. Horvath, S. Kondo, H. Matsui & H. Yoshimura, pp. 154–66. Tokyo: University of Tokyo.

Huizinga, J. (1972). Casual blood pressure in populations. In *Human Biology of Environmental Change*, ed. D. J. M. Vorster, pp. 164–9. London: International Biological Programme.

Hull, D. (1979). Migration, adaptation, and illness: a review. *Social Science and Medicine*, **13A**, 25–36.

Hulse, F. S. (1957). Exogamie et heterosis. *Archives Suisses d'Anthropologie Generale*, **22**, 103–25.

James, G. D., Jenner, D. A., Harrison, G. A. & Baker, P. T. (1985). Differences in catecholamine excretion rates, blood pressure and lifestyle among Western Samoan men. *Human Biology*, **57(4)**, 635–47.

James, W. H. (1966). The effect of altitude on fertility in Andean countries. *Population Studies*, **20**, 97–101.

Jenner, D. A., Reynolds, V. & Harrison, G. A. (1980). Catecholamine excretion rates and occupation. *Ergonomics*, **23**, 237–46.

210   M. A. Little & P. T. Baker

Johnston, F. E., Wainer, H., Thissen, D. & MacVean, R. (1976). Hereditary and environmental determinants of growth in height in a longitudinal sample of children and youth of Guatamalan and European ancestry. *American Journal of Physical Anthropology*, **44**(3), 469–76.

Joseph, J. G., Prior, I. A. M., Salmond, C. E. & Stanley, D. (1983). Elevation of systolic and diastolic blood pressure associated with migration: The Tokolau Island Migrant Study. *Journal of Chronic Diseases*, **36**(7), 507–16.

Kalla, A. K. (1983). Melanin pigmentation: its variation and genetics studies through Indian populations. *Indian Journal of Physical Anthropology and Human Genetics*, **9**(1–3), 171–98.

Kasl, S. V. & Berkman, L. (1983). Health consequences of the experience of migration. *Annual Review of Public Health*, **4**, 69–90.

Kim, Y. S. (1982). Growth status of Korean school children in Japan. *Annals of Human Biology*, **9**(5), 453–8.

Kondo, S. & Eto, M. (1975). Physical growth studies on Japanese-American children in comparison with native Japanese. In *Comparative Studies on Human Adaptability of Japanese, Caucasians and Japanese Americans, JIBP Synthesis, Human Adaptability*, vol. 1, ed. S. M. Horvath, S. Kondo, H. Matsui & H. Yoshimura, pp. 13–45. Tokyo: University of Tokyo Press.

Kondo, S. & Kobayashi, K. (1975). Microevolution and modernization of Japanese. In *Anthropological and Genetic Studies on the Japanese, JIBP Synthesis, Human Adaptability*, vol. 2, ed. S. Watanabe, S. Kondo & E. Matsunaga, pp. 5–14. Tokyo: University of Tokyo Press.

Lasker, G. W. (1946). Migration and physical differentiation. A comparison of immigrant and American-born Chinese. *American Journal of Physical Anthropology*, **4**, 273–300.

Lasker, G. W. (1952). Environmental growth factors and selective migration. *Human Biology*, **24**, 262–89.

Lasker, G. W. (1954). The question of physical selection of Mexican migrants to the USA. *Human Biology*, **26**, 52–8.

Lasker, G. W. (1969). Human biological adaptability. *Science*, **166**, 1480–6.

Lee, E. S. (1966). A theory of migration. *Demography*, **3**, 47–57.

Leslie, P. W. (1985). Potential mates analysis and the study of human population structure. *Yearbook of Physical Anthropology*, **28**, 53–78.

Levi, L. (ed.) (1972). *Stress and Distress in Response to Psychosocial Stimuli: Laboratory and Real-Life Studies on Sympatho-Adrenomedullary and Related Reactions*. Oxford: Pergamon Press.

Lieber, M. D. (ed.) (1977). *Exiles and Migrants in Oceania*. Honolulu: University Press of Hawaii.

Little, M. A. (1980). Designs for human-biological research among savanna pastoralists. In *Human Ecology in Savanna Environments*, ed. D. R. Harris, pp. 479–503. London: Academic Press.

Little, M. A. (1982). The development of ideas on human ecology and adaptation. In *A History of American Physical Anthropology, 1930–1980*, ed. F. Spencer, pp. 405–33. New York: Academic Press.

Little, M. A. & Baker, P. T. (1976). Environmental adaptations and perspectives. In *Man in the Andes: A Multidisciplinary Study of High-Altitude Quechua*, ed. P. T. Baker & M. A. Little, pp. 405–28. Stroudsburg, Penn.: Dowden, Hutchinson and Ross.

Little, M. A., Galvin, K., Shelley, K., Johnson, B. R., Jr, & Mugambi, M. (1987). Resources, biology and health of pastoralists. *Proceedings of the Conference: Arid Lands Today and Tomorrow.* Tucson: University of Arizona.

Little, M. A., Thomas, R. B., Mazess, R. B. & Baker, P. T. (1971). Population differences and developmental changes in extremity temperature responses to cold among Andean Indians. *Human Biology*, 43(1), 70–91.

Loomis, W. F. (1970). Rickets. *Scientific American*, 223(6), 77–91.

Lundberg, U. (1976). Urban commuting: crowdedness and catecholamine excretion. *Journal of Human Stress*, 2, 26–32.

Malina, R. M., Buschang, P. H., Aronson, W. L. & Selby, A. (1982). Childhood growth status of eventual migrants and sedentes in a rural Zapotec community in the Valley of Oaxaca, Mexico. *Human Biology*, 54(4), 709–16.

Malhotra, M. S., Ramaswamy, S. S. & Ray, S. V. (1960). Effect of environmental temperature on work and resting. *Journal of Applied Physiology*, 15, 769–70.

Martorell, R. (1980). Interrelationships between diet, infectious disease, and nutritional status. In *Social and Biological Predictors of Nutritional Status, Physical Growth, and Neurological Development*, ed. L. S. Greene & F. E. Johnston, pp. 81–106. New York: Academic Press.

Mascie-Taylor, C. G. N. (1984). The interaction between geographical and social mobility. In *Migration and Mobility: Biosocial Aspects of Human Movement*, ed. A. J. Boyce, pp. 161–78. London: Taylor and Francis.

Mason, E. D. & Jacob, M. (1972). Variations in basal metabolic rate responses to changes between tropical and temperate climates. *Human Biology*, 44(1), 141–72.

Mazess, R. B. & Baker, P. T. (1964). Diet of the Quechua Indians living at high altitude. *Journal of Clinical Nutrition*, 15, 41–351.

McGarvey, S. T. & Baker, P. T. (1979). The effects of modernization and migration on Samoan blood pressures. *Human Biology*, 51(4), 461–79.

Mills, C. A. (1942). Climatic effects on growth and development with particular reference to the effects of tropical residence. *American Anthropologist*, 44(1), 1–13.

Monge, C. (1948). *Acclimatization in the Andes: Historical Confirmations of 'Climatic Aggression' in the Development of Andean Man.* Baltimore: The Johns Hopkins Press.

Moran, E. (1981). *Developing the Amazon.* Bloomington: Indiana University Press.

Mueller, W. H., Schull, V. N., Schull, W. J., Soto, P. & Rothhammer, F. (1978a). A multinational Andean genetic and health program: growth and development in an hypoxic environment. *Annals of Human Biology*, 5, 329–52.

Mueller, W. H., Yen, F., Rothhammer, F. & Schull, W. J. (1978b). A multinational Andean genetic and health program: VI. Physiological measurements of lung function in an hypoxic environment. *Human Biology*, 50, 489–513.

Mueller, W. H., Yen, F., Rothhammer, F. & Schull, W. J. (1978c). A multinational Andean genetic and health program: VII. Lung function and physical growth – multivariate analyses in high- and low-altitude populations. *Aviation, Space and Environmental Medicine*, 49(10), 1180–96.

Mueller, W. H., Yen, F., Soto, P., Schull, V. N., Rothhammer, F. & Schull, W.

J. (1979). A multinational Andean genetic and health program: VII. Lung function changes with migration between altitudes. *American Journal of Physical Anthropology*, 51, 183–195.

Muller, W. H., Murillo, F., Palomino, H., Badzioch, M., Chakraborty, R., Fuerst, P. & Schull, W. J. (1980). The Aymara of western Bolivia: V. Growth and development in an hypoxic environment. *Human Biology*, 52(3), 529–46.

Mugambi, M. & Little, M. A. (1983). Blood pressure in nomadic Turkana pastoralists. *East African Medical Journal*, 60(12), 863–9.

Murra, J. (1972). El 'control vertical' de un maximo de pisos ecológicos en la economía de las sociedades Andinas. In *Visita de la Provincia de Leon de Hánuco (1562), Iñigo Ortiz de Zúñiga, Visitador*, vol. II. Huánuco, Perú: Universidad Hermillo Valdizen.

Neer, R. M. (1975). The evolutionary significance of vitamin D, skin pigment, and ultraviolet light. *American Journal of Physical Anthropology*, 43, 409–16.

Ortner, D. J. (ed.) (1983). *How Humans Adapt: A Biocultural Odyssey*. Washington, DC: Smithsonian Institution Press.

Parry, E. H. O. (1969). Ethiopian cardiovascular studies III. The casual blood pressure in Ethiopian highlanders in Addis Ababa. *East African Medical Journal*, 46(5), 246–56.

Pawson, I. G. & Janes, C. (1981). Massive obesity in a migrant Samoan population. *American Journal of Public Health*, 71, 508–13.

Pawson I. G. & Janes, C. (1982). Biocultural risks in longevity: Samoans in California. *Social Science and Medicine*, 16(2), 183–90.

Pelletier, D. L. & Hornick, C. A. (1987). Blood lipid studies. In *The Changing Samoans: Health and Behavior in Transition*, ed. P. T. Baker, J. M. Hanna & T. S. Baker. New York: Oxford University Press.

Picón-Reátegui, E. (1976). Nutrition. In *Man in the Andes: A Multidisciplinary Study of High-Altitude Quechua*, ed. P. T. Baker & M. A. Little, pp. 208–36. Stroudsburg, Penn.: Dowden, Hutchinson and Ross.

Pietrek, J., Windo, J., Preece, M. A., O'Riordan, J. L. H., Dunnigan, M. G., McIntosh, W. B. & Ford, J. A. (1976). Prevention of vitamin-D deficiency in Asians. *Lancet*, 1, 1145–8.

Preece, M. A., McIntosh, W. B., Tomlinson, S., Ford, J. A., Dunnigan, M. G., & O'Riordan, J. L. H. (1973). Vitamin-D deficiency among Asian immigrants to Britain. *Lancet*, 1, 907.

Prior, I. A. M. (1971). The price of civilization. *Nutrition Today*, Jul/Aug, 2–11.

Prior, I. A. M. (1981). The Tokelau Island Migrant Study: a progress report 1979. In *Migration, Adaptation and Health in the Pacific*, ed. C. Fleming & I. Prior, pp. 27–38. Wellington, New Zealand: Wellington Hospital Epidemiology Unit.

Prior, I. A., Davidson, F., Salmond, C. E. & Czochanska, Z. (1981). Cholesterol, coconuts and diet on Polynesian atolls; a natural experiment: the Pukapuka and Tokelau Island Studies. *American Journal of Clinical Nutrition*, 34, 1552–61.

Prior, I. A. M., Harvey, H. P. B., Neave, M. I. E. & Davidson, F. (1966). *The Health of Two Groups of Cook Island Maoris*. Special Report Series 26. Wellington, New Zealand: New Zealand Department of Health.

Prior, I. A. M., Hooper, A., Huntsman, J. W., Stanhope, J. M. & Salmond, C. E. (1977). The Tokelau Island migrant study. In *Population Structure and Human Variation*, ed. G. A. Harrison, pp. 165–86. Cambridge: Cambridge University Press.

Prior, I. A. M., Stanhope, J. M., Evans, J. G. & Salmond, C. E. (1974). The Tokelau Island Migrant Study. *International Journal of Epidemiology*, 3, 225–32.

Ramirez, M. E. & Mueller, W. H. (1980). The development of obesity and fat patterning in Tokelau children. *Human Biology*, 52, 675–87.

Reinhold, J. G., Lahimgarzadeh, A., Nasr, K. & Hedayati, H. (1973). Effects of purified phytate and phytate-rich bread upon metabolism of zinc, calcium, phosphorus, and nitrogen in man. *Lancet*, 1, 283–8.

Reynolds, V., Jenner, D. A., Palmer, C. D. & Harrison, G. A. (1981). Catecholamine excretion rates in relation to life-styles in the male population of Otmoor, Oxfordshire. *Annals of Human Biology*, 8, 197–209.

Riley, R. A. (1979). A dietary survey of downward Indian migrants and long-term coastal residents living in southern coastal Peru. *Archivos Latinoamericanos de Nutrición* (Caracas), 29(1), 69–102.

Roberts, D. F. (1952). Basal metabolism, race, and climate. *Journal of the Royal Anthropological Institute*, 82(2), 169–83.

Roberts, D. F. (1978). *Climate and Human Variability*, 2nd edn. Menlo Park, Cal.: Cummings Publishing.

Roberts, J. & Mauer, K. (1977). *Blood Pressure Levels of Persons 6–74 Years, United States, 1971–1974*. Publ. NO(HRA) 78–1648, Series 11, No. 203. Washington, DC: Department of Health, Education and Welfare.

Robins, R. S. (1981). Disease, political events and populations. In *Biocultural Aspects of Disease*, ed. H. R. Rothschild, pp. 153–75. New York: Academic Press.

Roe, D. A. (1985). *Drug-Induced Nutritional Deficiencies*, 2nd edn. Westport, Conn.: AVI Publishing.

Schull, W. J. & Rothhammer, F. (1977). A multinational Andean genetic and health programme: a study of adaptation to the hypoxia of altitude. In *Physiological Variation and its Genetic Basis*, ed. J. S. Weiner, pp. 139–69. London: Taylor and Francis.

Scotch, N. (1960). A preliminary report on the relation of sociocultural factors to hypertension among the Zulu. *Annals of the New York Academy of Science*, 84, 1000–9.

Scudder, T. (1980). River-basin development and local initiative in African savanna environments. In *Human Ecology in Savanna Environments*, ed. D. R. Harris, pp. 383–405. London: Academic Press.

Shaper, A. G., Leonard, P. J., Jones, K. W. & Jones, M. (1969). Environmental effects on the body build, blood pressure and blood chemistry of nomadic warriors serving in the army in Kenya. *East African Medical Journal*, 46(5), 282–9.

Shaper, A. G. & Saxton, G. A. (1969). Blood pressure and body build in a rural community in Uganda. *East African Medical Journal*, 46(5), 228–45.

Shapiro, H. H. (1939). *Migration and Environment: A Study of the Physical Characteristics of the Japanese Immigrants to Hawaii and the Effects of Environment on Their Descendants*. London: Oxford University Press.

214   M. A. Little & P. T. Baker

Siervogel, R. M. (1983). Genetic and familial factors in essential hypertension and related traits. *Yearbook of Physical Anthropology*, **26**, 37–63.
Skinner, B. F. (1981). Selection by consequences. *Science*, **213**, 501–4.
So, J. (1975). Genetic, acclimatizational and anthropometric factors in hand cooling among north and south Chinese. *American Journal of Physical Anthropology*, **43**(1), 31–8.
Spector, R. M. (1971). *Mortality Characteristics of a High-Altitude Peruvian Population*. MA Thesis in Anthropology. University Park, Pa.: Pennsylvania State University.
Stanhope, J. M., Prior, I. A. M. & Fleming, C. (1981). The Tokelau Island Migrant Study: morbidity of adults. In: *Migration, Adaptation and Health in the Pacific*, ed. C. Fleming & I. Prior, pp. 44–55. Wellington, New Zealand: Wellington Hospital Epidemiology Unit.
Steegmann, A. T., Jr (1974). Ethnic and anthropometric factors in finger cooling: Japanese and Europeans in Hawaii. *Human Biology*, **46**, 621–31.
Steward, J. H. & Faron, L. C. (1959). *Native Peoples of South America*. New York: McGraw-Hill.
Susanne, C. (1984). Biological differences between migrants and non-migrants. In *Migration and Mobility: Biosocial Aspects of Human Movement*, ed. A. J. Boyce, pp. 179–93. London: Taylor and Francis.
Swan, C. H. & Cooke, W. T. (1971). Nutritional osteomalacia in immigrants in an urban community. *Lancet*, **2**, 456–9.
Taylor, R. & Zimmet, P. (1981). The influence of variation in obesity in the difference in the prevalence of abnormal glucose tolerance in Tuvalu. *New Zealand Medical Journal*, **94**, 176–8.
Thieme, F. P. (1957). A comparison of Puerto Rico migrants and sedentes. *Michigan Academy of Sciences, Arts and Letters*, **42**, 249–67.
Thomas, D. B. (1979). Epidemiological studies of cancer in minority groups in the western United States. *National Cancer Institute Monograph*, **53**, 103–13.
Ursin, H., Baade, E. & Levine, S. (eds) (1978). *Psychobiology of Stress: A Study of Coping Men*. New York: Academic Press.
Vallery-Masson, J., Bourlier, F. & Poitrenaud, J. (1980). Can a protracted stay in the tropics permanently lower basal metabolic rates of European expatriates? *Annals of Human Biology*, **7**(3), 267–71.
Vaughan, J. P. & Miall, W. E. (1979). A comparison of cardiovascular measurements in the Gambia, Jamaica and the Republic of Tanzania. *Bulletin of the World Health Organization*, **57**, 281–9.
Ward, R. H. (1983). Genetic and sociocultural components of high blood pressure. *American Journal of Physical Anthropology*, **62**, 91–105.
Way, A. B. (1976a). Morbidity and postneonatal mortality. In *Man in the Andes: A Multidisciplinary Study of High-Altitude Quechua*, ed. P. T. Baker & M. A. Little, pp. 147–60. Stroudsburg, Penn.: Dowden, Hutchinson and Ross.
Way, A. B. (1976b). Exercise capacity of high-altitude Peruvian Quechua Indians migrant to low altitude. *Human Biology*, **48**(1), 175–91.
Weil, C. (1979). Morbidity, mortality and diet as indicators of physical and economic adaptation among Bolivian migrants. *Social Science and Medicine*, **13D**, 215–22.
Weiss, K. M., Ferrell, R. E. & Hanis, C. L. (1984). A New World Syndrome of

metabolic diseases with a genetic and evolutionary basis. *Yearbook of Physical Anthropology*, **27**, 153–78.

Wheeler, E. & Tan, S. P. (1983). Trends in the growth of ethnic Chinese children living in London. *Annals of Human Biology*, **10(5)**, 441–6.

Wilkie, J. W. & Perkal, A. (eds) (1984). *Statistical Abstract of Latin America*, vol. 23. UCLA Latin American Center Publications. Los Angeles: University of California.

Yoshimura, H. & Horvath, S. M. (1975). Comparative studies on the response to cold of Japanese and Caucasians. In *Comparative Studies on Human Adaptability of Japanese, Caucasians and Japanese Americans, JIBP Synthesis, Human Adaptability*, vol. 1, ed. S. M. Horvath, S. Kondo, H. Matsui & H. Yoshimura, pp. 167–75. Tokyo: University of Tokyo Press.

Yoshimura, H. & Iida, T. (1952). Studies of the reactivity of skin vessels to extreme cold. Part II. Factors concerning the individual difference of the reactivity, or the resistance against frostbite. *Japanese Journal of Physiology*, **2(2)**, 177–85.

Zimmet, P. (1982). Type 2 (non-insulin-dependent) diabetes – an epidemiological overview. *Diabetologia*, **22(6)**, 399–411.

Zimmet, P. & Björntorp, P. (1979). Adipose tissue cellularity in obese non-diabetic men in an urbanized Pacific Island (Polynesian) population. *American Journal of Clinical Nutrition*, **32**, 1788–91.

Zimmet, P., Taft, P., Guinea, A., Guthrie, W. & Thoma, K. (1977). The high prevalence of diabetes mellitus on a central Pacific island. *Diabetologia*, **13**, 111–15.

8 Migration and disease

BERNICE A. KAPLAN

Two things seem certain about humans wherever they have lived: they have been reproductively successful; and they have been plagued by diseases and other ill health, which, until recently, has kept their numbers in check. As humans became a numerically successful species they began to expand: first, nearby, and later, into more distant environments away from their natal territories. In so doing they met new disease vectors in the new environments and they also brought many of their 'native diseases' with them. As we examine the role of migration as a factor in the dissemination of disease, there is no claim to all inclusiveness; we go no further back in history than to the thirteenth century. Most of the emphasis in this chapter will be on the role of migration in the spread and acquisition of disease during the nineteenth and twentieth centuries. The instances cited are examples of the influences migration has had on disease and do not necessarily affect more people than other diseases not discussed.

Migration as a source for disease transmission

People move from place to place for many reasons. Today there are relatively few traditional nomadic populations, but temporary migrant labour still forms a large portion of the transitory agricultural work force in many parts of the world. Other, more permanent, labour migration, both intranational and international, has been an important source for new workers. Many parts of the Third World have sent labourers to the industrialized West. North Africa, Yemen, Turkey, India, Pakistan and East Africa have been sources for many non-Europeans now in Germany, France, Switzerland, Great Britain and Scandinavia, while workers (documented or not) from these countries and from countries in Middle, Central, and South America are found in North America. People on overcrowded islands in the Pacific are also on the move, to New Zealand and Australia, Hawaii and the United States mainland. For many of these populations there is good documentation on the effect of migration on their health. As the moves of such labourers involve adaptations to a

216

lifestyle more consonant with Western industrialized societies, marked by increased urbanization and a move away from rural peasant social structures (see Bogin, this volume), researchers have found evidence for a growing prevalence of so-called diseases of Westernization (obesity, hypertension, diabetes, coronary heart disease, mental illness). The extent of these disabilities in various populations will be considered. In the last several decades widespread famine has become an important motivation for human migration (in the Sahel, in the Tigre region of Ethiopia, and now, increasingly, famine has become a stimulus for migration in the Sudan). The disease experiences of these starving populations on the move can be compared with the experiences of those fleeing from the famines caused by the potato blight in Ireland during the last century.

Often diseases were introduced accidentally into areas where they were unknown before. Refugees fleeing religious persecution left England, Germany and Russia for the New World. Others left France for similar reasons and settled in England. From the Western world Christian missionaries have proliferated in the less developed world, challenging not only the host countries' belief system, but their marriage, kinship, inheritance, educational and political systems as well. In the wake of the religious missionary came the support system for the new faith (the doctors, teachers, lawyers, economic investors and entrepreneurs, police and other armed forces) all of whom brought with them to new regions their national ideas about health and disease, again replacing the traditional ideas of the missionized people. The new health practices they introduced succeeded in keeping alive more babies, more mothers (to make more babies) and more elderly to be supported by the mothers and their grown children and, to all of them, along with changed diets and different ways of earning a livelihood, they introduced the diseases of Western society.

Early religious zealots travelled east toward the 'Holy Land' Jerusalem to convert the infidel and came back, defeated and infested. Their eastward forays coincided with a widespread growth of coastal and overland trade routes. Traders and crusaders brought back the rat and his flea and spread the Black Death over all Europe and North Africa. Massive pilgrimages today, not only to Mecca but also to shrines important to other faiths, pose sanitation and public health problems to the host locales.

The institution of economic slavery in the eighteenth and nineteenth centuries saw the forced removal (migration) of millions of Africans to North and South America where the trepanoma-borne yaws was commonly found in areas populated by African slaves and where the sickle-cell trait and associated anaemia were also newly introduced. Probably malaria was also transported with these populations. When slave trading ceased in North America the incidence of yaws declined dramatically as the North American climate was not conducive to its continuation without a source for reinfection (Cowan-Ricks, personal communication; Hopkins & Florez, 1977, pp. 349–55; Savitt, 1978, pp. 76–7).

Infectious diseases and migration

Massive population increases result in large-scale population redistribution (both rural and urban) and much of this movement is not matched by equal expansion of available resources (housing, water supply, sewage and water disposal). Taken together with difficulties in earning a livelihood in new locales, these factors lead to heavy pressure on the physical and mental health of the migrants (Prothero, 1977). Population increase also leads to heavy exploitation of the land when the production of crops is meant for international exchange rather than for local consumption. Over-intensive cultivation of the land, whether for crops or for forage, leads to atmospheric changes which are, in part, responsible for droughts, desertification, famine and further forced movement of the population. And, if this were not enough, the natural and man-made disasters promote the migration of ever-increasing numbers of people. In addition, the movements of those forced to emigrate as the result of wars or political disruption may lead to the spread of communicable diseases across national boundaries to the detriment of all.

   As a result of increasingly sophisticated means of transportation, the possibilities for the rapid spread of communicable disease have grown astronomically. In modern times what has, so far, prevented the spread of plagues such as the Black Death has been a combination of sanitary practices: sewage disposal and treatment; purification of water supplies; cultural attitudes regarding cleanliness of place, person, and comestibles; readily available laundry facilities; and garbage disposal. Furthermore, the poorest, who are most likely to be at risk from infections in most populations, have lacked the means to make long and speedy journeys. Lengthy quarantine regulations exist in some countries for animals and plants being introduced from abroad, but no longer are travellers, immigrants or other foreigners detained and medically examined before being given entry permits to their destinations. Immigration quotas and the necessity for visas today serve as partial limiters to free passage, but such regulators are directed more against selected populations than as protection against the importation of disease.

LeRiche, Harding & Milner (1971) note the great fear of the authorities that infectious diseases will be spread through air travel, especially those diseases which are quarantinable but which may slip through: cholera, plague, smallpox, louse-borne typhus, yellow fever. Other epidemic diseases could also fly in on the wings of a jet plane (or in the clothing of the migrants): meningitis, dengue fever, malaria, influenzas, poliomyelitis, trachoma and trepanosomiasis. Clearly migrants are responsible not only for adapting to the diseases in the environments to which they move, but also for bringing diseases with them.

Both hereditary and communicable diseases are spread by carriers, often enough by people who host the agent without being themselves obviously affected. Typhoid Mary comes to mind or, as recently as June 1985, another example was the young European hitchhiker who returned to Great Britain by air from the Indian subcontinent and who spent several nights in youth hostels before being hospitalized and diagnosed as having typhoid. In the intervening period he had unknowingly exposed those with whom he had been in contact. In much the same manner the Spanish Conquistadors in the sixteenth century introduced smallpox and the so-called childhood diseases to the New World with disastrous results, as people with no natural resistance to these scourges were decimated by succeeding epidemic waves of measles, whooping cough and chickenpox. If the relationship between migration and disease is examined only in terms of the individual who migrates, it would be possible to miss both the long-range effects of the migrants on the population structure of the host area and the long-range effects of the environment on the population introduced from afar. It is precisely because a multigenerational model was used by Livingstone (1958) that the tie between human modification of the environment by migrants occupying new regions and the genetic adaptation to malaria was documented. Some changes occur rapidly enough to be observed as they happen (e.g. the European introduction of infectious diseases to the populations they encounter in the Americas). Others, such as the diseases associated with modernization, change their rates more slowly and are best monitored in different ethnic groups for several generations.

Territorial expansion, mobility and disease

We will not be considering the use of germ warfare in these pages, but should note that the colonist European population while on their west-ward trek to 'conquer the continent' of North America saw no wrong in trading smallpox-infested blankets to the native Amerind population as a means of reducing the threat of native opposition to their takeover. The transmission of disease through trade is a two-way activity, however, and while the Europeans were spreading smallpox and encouraging alcoholism among the native populations (by trading in 'firewater'), the Indians were sharing their sacred smokes in solemn recognition of treaties agreed to, consuming tobacco which they used sparingly in religious ceremony but to which the Europeans soon became addicted. Whereas the damage done centuries ago by smallpox infestations is no longer a threat to the surviving Amerinds, the Indian drug brought lung cancer and other diseases, including coronary heart disease, which are associated with the inhalation of tobacco smoke, and which have come to be thought of as Western diseases.

Short-duration tourism must also be mentioned as a source for disease experiences that would not occur in more sedentary societies. Amoebic and other dysenteries often accompany returning tourists, as do some of the more exotic diseases (dengue fever, hepatitis, etc.). On occasion these same travellers are the source of introduction of new influenzas, other non-local contagious diseases or even the introduction of new potential plagues; for example, the first known case and death from AIDS (acquired immune deficiency disease) in China was reported in a visiting tourist in The Times, 30 July 1985.

In some instances migrants, because of their need to adapt to changed environmental conditions, have so modified the environment that they made possible the establishment of diseases hitherto unknown in the region. Livingstone (1958) traced the migration of populations to West Africa where rice and yam agriculture was introduced, and where clearing of the land for seeding left behind stagnant pools in which mosquitoes could reproduce. Soon thereafter malaria became endemic in the population and, in the intervening generations, selection led to the establishment of the polymorphic sickle-cell trait as a human adaptation to the disease threat.

Industrial diseases and migrant workers

Almost everyone in the modern Western world is a migrant, moving for the chance for a job, or for a promotion up the career ladder, or to a dream retirement village. Each move brings possibilities for exposure to new diseases, just as each migrant brings the burden of the exposures already acquired in previous residences. In new environments adaptations suitable in earlier settings are often no longer effective. Sometimes the new work environment poses threats unthought of when the decision to move was made. More often than not the necessity for employment is sufficiently pressing that environmental risks are downplayed or denied until irreversible damage has been done. Some of the most pressing health problems of the present day are tied to new diseases and to significant increases in the rates of occurrence of industrially borne diseases, where the proof of connection between the industry and the disease it spawns remains disguised behind an average two-decade incubation period.

Migrant studies do not focus on the new industrial diseases. To uncover them in their full extent would threaten the basis of the post World War II chemically modified world. To close the plants creating both nuclear energy and slowly decaying nuclear waste would severely limit the possibilities opened by the energy-dependent electronic revolution. Such a move could potentially threaten the growing energy needs of a 5-billion-plus hominid population. It is difficult to imagine the extent of the ensuing unemployment and poverty-based social and political upheaval. So the pollution encountered by new migrants and sedentes alike will continue, new diseases and genetic anomalies will proliferate until a crisis point is reached. Only then will serious research directed at prevention and cure finally begin. In the meantime, researchers' attention is principally directed to old diseases which still threaten both the new migrant, the extant population in the receiving areas, and often the sedentes left at home as well.

Migrant study design and methodological problems

In all instances of migration, changed lifestyles lead to differential occurrences of disease. Although convincing arguments can be made for many, if not most, of the suggested causative factors, it is often difficult to design studies which completely control the variables. Thus, migrants may enter dramatically different environments which vary from their home locales not only in climate, foods available, work and leisure activities, housing and sanitation facilities, expected cultural behaviours and social relations, crowded housing, language differences, varying ethnic prejudices etc., but even in clothing, patterns of bodily movement and access to help when needed.

If we look at changing disease patterns in migrant populations, differences can be documented between disease experiences of the non-migrant sedentes, the permanent migrants, those who returned to their natal places, the offspring of migrants in the new locales and the non-migrant natives of the new locales. To date no study has taken into consideration simultaneously all five groups, although a variety of studies have considered in various combinations each of these populations. Nor has any study yet attempted to account for all the variables encountered by migrants in their new settings. In this regard we should recognize that every study, even the best, is afflicted with what might be called 'inborn errors of assumptions' so some will focus on diet, others on changes in altitude or duration of exposure to the new setting. Some will compare migrants and native residents; or migrants and those they left behind; or migrants, sedentes and returnees; or recent and older migrants; or will choose instead to focus on the different locales in which migrants find themselves: rural–urban, urban–urban or rural–rural. Studies based on all of these models yield interesting and statistically significant results, and in many instances the findings are strikingly parallel. What is not parallel, however, are the assumptions which underlie the studies and this often makes difficult the drawing of conclusions from comparisons of studies with differing emphases, although the direction towards which many studies point is clear enough. It has been observed (MacMahon & Pugh, 1970) that migrant populations carry their disease risks with them, but their offspring have disease risks characteristic of the host community.

Several methodological problems continue to plague migration studies. There still remains the question of whether the migrants are inherently different from the sedentes. Is a different segment of the population selected for migration? Are they younger, stronger, more alert, of a different socioeconomic status from those who remain behind? If so, can the migrants be shown to represent either a different biological population or one which in their native locale had made adaptations to a biological and sociocultural environment different from that of the sedentes?

Most of the studies comparing differences in disease rates between migrants and sedentes look at these populations at a single point in time (Phelps, Sobel & Fisher, 1961; Smith *et al.*, 1964; Stromberg, Peyman & Dowd, 1974; Nadim, Amini & Malek-Afzali, 1978; Cochrane & Stopes-Roe, 1981; Slesinger & Cautley, 1981; Sornmani *et al.*, 1983; Gerber, 1984), but three are longitudinal studies where the migrants and their descendents have been followed for periods up to two and three decades. The Tokelau Island Migrant Study in Polynesia, has traced biomedical changes in migrant, sedente and returned populations for over 20 years (Prior & Tasman-Jones, 1981). Also in Polynesia, United States and Australian scientists have been studying Samoan migrants (from Western Samoa to Australia and from United States Samoa to Hawaii and the US mainland) (Prior & Tasman-Jones, 1981; Little & Baker, this volume). These studies have been designed to control for environment, diet, age, occupation and health experiences. Another longitudinal study has focused on adaptation of high-altitude Peruvian populations to low altitude as compared to sedente populations already living at low altitude (Little & Baker, this volume). Many of the results of this investigation come from questionnaires to which the subjects responded.

Migration and the expansion of Western diseases

Many of the diseases of Westernization are considered to be stress-related, though here again there is little agreement as to what causes stress. Discrepancies between early learned and newly expected sociocultural responses lead to anxiety often associated with stress. So also do overcrowding, unfamiliar work conditions, unstable family relations and uncertainty about employment. Furthermore, different people respond to stress in varying ways: for some, bizarre behaviour signals distress; others have gastrointestinal disorders; for some, food binges may relieve anxiety but lead to obesity and adult-onset diabetes; dietary problems may lead to coronary problems; unhealthy environments may exacerbate the development of malignancies, or cause spontaneous abortions or genetic defects.

In virtually all migrant studies changes have been observed in diet, in salt intake, in body weight and in exercise levels as compared with the pre-migrant experiences. It is often assumed that a better diet is a measure of better economic conditions. Yet if we consider the growth of 'diseases of Westernization' endemic in the modernized world, we might well question – assuming that the oft-cited relationship between diet and disease is correct – whether more is better. Calorific intake is higher, but so also are fat, sugar and salt consumption and, following on this, obesity has come to be characteristic not only of migrants to Western society, but also of many of the birth-right nationals. Obesity among the migrants as reported in the following studies is the more marked because in most instances it was not as characteristic of the populations before migration.

There appears to be a relationship between disease risks which confront migrants to places characterized by modernization, Westernization, and industrialization, whether these features are found in settings of large-scale agriculture (agrobusinesses) or in more urbanized settings. Thus, there are dietary changes, activity load changes, as well as early onset of obesity associated with an increased risk of hypertension, atherosclerosis, coronary heart disease and diabetes in populations resident in the industrialized world. Differing environmental risks are also manifest in cancer rates which differ between sedente populations and migrants to the new locale. We will look at some of these increased disease risks of migration.

Hypertension

When migrants are compared with sedentes, the migrants show more hypertension. In the Tokelau study of the inhabitants of a Western Pacific tropical island there and in New Zealand, for example, sedente and migrant populations show some reduction in weight with age as well as an increase in blood pressure. The migrants, however, were found to have blood pressure rates more than one standard deviation higher than that of the sedentes (Prior & Tasman-Jones, 1981). In this population the children also showed systolic pressure significantly higher than that of the Tokelau sedente children. This was true of migrant girls up to the age of eight and of migrant boys of all ages. In the boys the higher systolic rate was associated with weights which were significantly higher in New Zealand than were those of Tokelau youngsters.

The Asian (Indian–Pakistani) population in South Africa today represents third and fourth generation descendants of migrants who were brought to South Africa at about the beginning of the twentieth century to work on sugar plantations. They have been studied in South Africa although no direct comparisons have been made with those remaining in the areas of origin. Nonetheless it has been reported that this population shows a higher prevalence of hypertension than in Indians in India or in many White populations (Walker, 1981). There was more hypertension among the adult women (22%) than among adult men (15%), although these percentages are less than those reported of middle-age obesity.

In South Africa comparison has been made of rural and urban Blacks, where it is understood that the urban Blacks have migrated from rural areas, but again no systematic migrant–sedente study is reported. Nonetheless, there is considerable evidence to support the statement that throughout Africa there is a rural–urban difference in blood pressure which in urban South Africa is matched by high levels of obesity (Walker, 1981).

Among migrants to Israel, Modan (1981) notes a somewhat higher prevalence of hypertension in those born in North Africa and a lower one among the Yemenites. Associated with these findings is a higher rate of severe kidney disease and of death from stroke in those born in North Africa.

A migrant–sedente study conducted in Iran was designed to test the impact of urban living on the migrants. The sedente population in north-west Iran was compared with former residents who had moved from there to Teheran. The latter, in turn, were compared with the established urban residents in Teheran (Nadim *et al.*, 1978). There are two sedente populations in this study: one rural and one urban. For all ages and both sexes the urban migrants had blood pressures higher than those of the rural sedentes, but when migrants to the city were compared with the sedente urban population there was not much difference between the two groups.

As in the South African studies, Nadim *et al.* (1978) report that Iranian women have a higher prevalence of hypertension than men which holds both for the rural sample and for the migrants in the urban sample. The authors report an association between body build (obesity) and hypertension, but they are not convinced that different degrees of obesity are sufficient to explain the differences observed between the urban and rural samples, because within the obese groups the rural–urban differences are still found for all age and sex groups. Nadim *et al.* (1978) rule out as potentially associated factors: cigarette smoking, occupation, education and salt intake. They also note, without claiming as an association with hypertension, that migrants face many urban problems not confronting the sedente rural population: pollution, crowding, noise, high rents, inflation, traffic problems, insufficient public services and the need to learn another language (the migrants are Turkish speakers, but Farsi is the language of Teheran).

Nadim *et al.* (1978) also cite other studies showing an association between migration to urban areas and higher levels of hypertension for Zulus in Durban (Scotch, 1963), Easter Island migrants to mainland Chile (Cruz-Coke, Etcheverry & Nagel, 1964), rural–urban migrants in Guatemala (Hoobler *et al.*, 1965) and, in 1973, a study of migration of high-altitude-dwelling Aymara to Westernized lowlands in Chile (Cruz-Coke, Donoso & Barrera, 1973).

There seems little doubt that high blood pressure and migration experiences are associated. The extent to which the causes are psychological, nutritional or sociological, or are associated with a longer life expectancy in Western settings than in the place of origin is not yet fully clarified.

Contrary to findings in many other studies, the low blood pressures reported for high-altitude Peruvians remain low after migration to low altitudes, and these findings persist both among migrants to the Tambo Valley in Peru, where the setting is primarily agricultural, and to the city of Mollendo (Table 8.1). Even where migrants lived at low altitude for a long period and have adopted the higher fat and lower carbohydrate diets of the low-altitude sedente population, there is no increase in the risk for cardiovascular diseases in this population.

226 B. A. Kaplan

Table 8.1. *Comparison of migrants from high and low altitude with sedentes in Peru*

	Low-altitude sedente compared with low-altitude migrant	Low-altitude sedente compared with high-to-low altitude migrant	High altitude sedente compared with high-to-low altitude migrant
Blood pressure	similar	greater	similar
Dietary protein content	similar	similar	similar
Accidental death rate	greater	much greater	not compared
Respiratory death rate	greater	much greater	not compared
Cardiovascular death rate	similar	less	not compared

(After Baker & Beall, 1982, p. 91.)

Atherosclerosis and coronary heart disease

Japanese who emigrated to the United States between 1890 and 1924 showed different patterns in causes of death from those who remained behind. Thus, in Japan, death from atherosclerotic heart disease is only 25% of the frequency of that in Japanese in the United States while deaths from stroke (intracranial vascular lesions) in Japan are between two and three times as common as in Japanese in the United States (MacMahon & Pugh, 1970).

Supportive of the notion that migrants carry the disease risks of their homeland with them, Modan (1981) reports that in Israel ischaemic heart disease continues to have a high incidence among European and American-born migrants, whereas it is low in Yemenites and intermediate among migrants from the rest of Asia and North Africa.

In South Africa, both the Asian Indian population and the Black population show rates of heart disease different from those of the Whites. The Indians arrived in South Africa early in the twentieth century, but the Whites are also migrants, primarily from the United Kingdom and the Netherlands. The rates of death from heart disease among the sedente (albeit urbanized) Blacks, the South African Asian Indians and the Whites show a gradient (Table 8.2).

A similar gradient is seen in the figures for life expectancy at birth, here shown with comparative figures for Blacks in New York City and for Whites in the United Kingdom (Table 8.3). Despite these figures, Walker (1981) observed that middle-aged Blacks live longer than middle-aged

Table 8.2. *Percentage of South Africans who die from certain specific causes*

Cause of death	Blacks	Asian Indians	Whites
Stroke	3–7	5	7–10
Ischaemic heart disease	7–8	13	30–31

(After Walker, 1981, pp. 288, 297.)

Whites, while among individuals aged 50 or more in 1950, only 22.2% of Indians survived to 1970, compared with 32.7% of Whites and 39.5% of Blacks.

The conclusion from these figures may be that Blacks and Asians in South Africa are less prone to heart disease than are the Europeans, although overall their longevity is well below that of the European population. While the South African Indians have a low death rate from coronary heart disease (CHD), Walker (1981) observes that in every country to which Asians have migrated their likelihood of experiencing coronary heart disease has shown a great increase. Of those Indians and Blacks who die from CHD, over 75% of the deaths occur before the age of 60, while in the European populations 66% of CHD deaths occur after the age of 60. Such mortality rates clearly contribute to the differential life expectancies in these populations. In addition it is assumed that these populations are also succumbing to diseases other than CHD and that some of the differences in life expectancy are related to a differential access to appropriate medical attention.

For those living in the Hawaiian Islands the half-century between 1910 and 1960 was marked by a dramatic increase in CHD, rising in that period

Table 8.3. *Life expectancy at birth*

	Life expectancy (years)				
	South African Blacks	New York City Blacks	South African Asian Indians	South African Whites	United Kingdom Whites
Male	51.2	61.2	59.3	64.5	69.2
Female	58.9	68.9	63.9	72.3	75.6

(After Walker, 1981, p. 200.)

Table 8.4. *Age-adjusted coronary heart disease mortality rates per 100 000 males in Hawaii*

Year	Japanese	Filipinos	Chinese	Caucasians	Hawaiians and part Hawaiians
1950	78	80	142	280	320
1970	160	236	194	360	380

(After Glober & Stemmerman, 1981, p. 323.)

from 2.9 per 100 000 to 105.0 per 100 000 (Table 8.4). It seems clear that modernization and urbanization were taking their toll. When Japanese migrants to Hawaii were compared with those who remained at home and with those who had migrated to California, a gradient similar to that observed in other studies was reported: the Japanese males in Hawaii experienced CHD at twice the rate reported for Japanese sedentes, while those in California had a CHD rate nearly 50% higher again than that of the Hawaiian-Japanese. The Hawaiian-Japanese who developed CHD were heavier than the sedentes in Japan, but not as heavy as the sedente Hawaiians. They also differed from those in Japan in having higher serum cholesterol, triglycerides, uric acid and glucose levels. Both in Japan and in Hawaii men with CHD had significantly high blood pressure. From these figures it appears that the Hawaiian sedentes are most at risk, perhaps because they have lived for more generations in this presumed 'island paradise', and surely because they are more obese than the other populations. They are also among the poorest residents on the Hawaiian Islands today and therefore may be assumed to be under greater stress than members of other populations.

Diabetes among migrants

In New York City from 1930 to 1960 susceptibility to diabetes was commonly held to be, at least in part, genetically transmitted and most common in the Jewish population. There was no connection made then between migration and the disease, nor to any 'thrifty genotype' hypothesis about genetic adaptation to starvation (Neel, 1962), nor even to obesity.

Today many more populations are being diagnosed as having diabetes, and the number of those determined to be non-insulin-dependent diabetics are also rising dramatically. The disease is global in its distribution, and has been documented as more highly associated with migrant than sedente populations among Chinese in New York, Hispanics (mostly Mexicans) in Texas and south-western United States, Samoans in Hawaii and California, in Tokulauan Polynesian migrants to New Zealand, and among Whites, Blacks and Asians in South Africa. In Latin America the reported frequency of diabetes is low.

Both insulin-dependent and non-insulin-dependent diabetes have shown marked increases during the present century in places where social and technical changes have been most marked. Insulin-dependent diabetes increased three times as much in urban as in rural Melanesia, Micronesia and Polynesia. It also increased in migrants from these islands to industrialized countries, as even higher rates of increase were observed in New Zealand and the United States (E. Bailey, personal communication). Non-insulin-dependent diabetes usually develops after age 40 and is usually associated with obesity.

In their study of Samoan migrants, Baker and his associates sought to test the commonly held hypothesis that with modernization there is an increase in obesity and in the degenerative diseases of middle age: diabetes, hypertension and cardiovascular disease (Baker, 1981). The small Samoan population seemed ideal because they shared a common genetic and cultural heritage and had experienced relatively rapid modernization on their native homeland on the Samoan Islands and in both migrant destinations, Hawaii and California. These parallel experiences made possible further comparisons between those who stayed put and those who settled abroad in varying cultural and physical environmental settings. Similar trends in the distribution of these diseases are reported for New Zealand, Oahu, in the urban areas of the United States as well as in United States Samoa and Western Samoa.

One striking characteristic of all Samoans is their extreme obesity, which is even more marked among the migrants than among the sedentes. In United States Samoa, a gradient is found such that in the most isolated and traditional of the villages, where there are no main roads linking settlements, the men and women are more obese than are people in villages located on the main road or in the towns near the most modern harbours. Among the men weight and fatness were closely associated with occupation (Baker, 1981).

It is also possible to trace the extent of development of diseases of modernization under the stimulus of migration among the Tokulauan migrants to New Zealand. By comparing the incidence of diabetes among sedente and migrant Tokulauans with the Maori in New Zealand the impact of exposure to a Westernized sociocultural environment is seen to

Table 8.5. *Percentage of middle-aged South Africans who are 20% or more heavier than 'ideal weight for height'*

	Males	Females
Whites	about 35	about 30
Blacks	10–20	40–60
Asian Indians	35–45	50–60

(After Walker, 1981, pp. 292, 300, 306.)

be a significant variable (Prior & Tasman-Jones, 1981). Among the women of these three populations the Maori rates are the highest, Tokulauan migrant rates are next and the Tokulauan sedentes lowest, but in all three groups the women had higher rates of diabetes than did the men. However, sex differences in rates are less marked among the New Zealand Maori and the incidence of diabetes increases with body mass (obesity) in both males and females. In addition, diabetes and striking obesity are more prevalent among the Maori than in other groups.

The occurrence of diabetes in South African migrant populations mirrors that reported in Polynesian migrants. Among middle-aged South African Asian Indians obesity is very common: 35–45% in males and 50–60% in females (Walker, 1981) (Table 8.5). Given such high levels of obesity it is not surprising that some reports indicate up to 25% of the Asian Indians over 50 in South Africa show abnormal glucose tolerance (Walker, 1966, cited by Walker, 1981). In India, high frequencies of diabetes are found in upper income groups in urban areas, but it is less common in rural areas, where 90% of the population live.

The incidence of diabetes among Blacks in urban South Africa (largely a migrant population, since permanent settlement in urban South Africa is not condoned by the White authorities) is about the same as it is among Whites. Even in rural areas diabetes is increasing (Walker, 1981), suggesting that determining factors are neither the migrant experience, nor characteristics of the migrant population, but are instead changes in the dietary experience of these populations which lead to an increased inability to tolerate glucose. Increased obesity is clearly a risk factor for all population groups.

Among the Asian Indians the disease prevalance increases with movement from rural to urban settings within India, and from continent to continent by migrants.

Chinese migrants to New York City (Gerber, 1984) show an association between reduced physical activity and unrestricted availability of food which is matched with an increased occurrence of diabetes. The New York lifestyle is associated with a more technologically advanced distribution system which leads the residents to more affluent consumption patterns. Migrants experience these changes in lifestyle most intensely, and have the highest incidences of diabetes and other diseases associated with overeating. Among Chinese in New York City the foreign born have the highest proportion of diabetic deaths, a proportion which increases with longer residence in the city. Chinese migrants to Hawaii also have a greater diabetic mortality rate than do the Hawaiian-born Chinese (Sloan, 1963).

Gastrointestinal diseases

In recent years considerable attention has been given to inadequacies in the diets of those living in industrial societies. The prevalence of junk foods, high in fats and sugars and low in bulk, leads to malnourishment in societal settings where an abundance of appropriate foods is available, but where fast foods are preferred to those requiring time for preparation. Such diets are seen as culprits in the development of obesity, diabetes, hypertension, coronary heart disease and gastrointestinal disorders. Because it is less easy to measure and quantify the effects of emotional stress, and possibly because emotional stress has been perceived as a female rather than a male problem, less emphasis is given in the literature to this putative causative factor of gastrointestinal disturbances. Thus, while it is well known that social and psychological anxiety in situations of uncertainty can lead to malfunction of the gastrointestinal system, none of the studies dealing with the interaction between the experiences of migration and the functioning of the gut relates to anything other than diet as a factor.

These gastrointestinal diseases lead less often to death than do diabetes, hypertension and CHD which may explain why there is less emphasis placed on them in migrant studies than on other diseases of modernization. Hiatus hernia, gallstones, varicose veins, atherosclerosis and diverticulosis have been found to be related. Among Japanese migrants to Hawaii, 52% were found on autopsy to have had diverticulosis, compared with only 1% of Japanese sedentes (Glober & Stemmerman, 1981). Among Filipino male migrants, on the other hand, where CHD, which is associated with atherosclerosis, had shown a very rapid rise, the diverticulosis rate remains low (Glober & Stemmerman, 1981). The Polynesian migrants to New Zealand have a lower frequency of diverticular disease than would be expected, and South African Asian Indians also show lower than expected rates for all gastrointestinal diseases (ulcerative colitis, diverticular disease, duodenal ulcer, hiatus hernia, gallstones and kidney stones) although in the latter population there is a very slow rate of increase.

Migration sites and cancer sites

Technological and industrial innovation has brought with it many more potentially carcinogenic risk factors as modern chemistry makes possible the production of pollutants more toxic than any known heretofore: dioxins, PCBs, PVCs, formaldehyde, the processing of asbestos; the insecticides, fertilizers and fumigants used in agriculture; lead in paint and gasoline (petrol); the food additives included in ready-to-eat quick foods; and all the byproducts of the harnessing of nuclear power. The human race has clearly infested its environment with new lethal products at the same time that the older scourges are being, or have been, brought under control. It is such polluted modern environments that the great majority of labour migrants enter when they leave their less well-developed homelands. Whatever their ultimate destination, they are certain to be exposed to more harmful toxic substances than they ever knew at home. It is easy to be carried away with the hypothesis that migrant diseases are a byproduct of a more affluent life style. One should not forget, however, that affluence produces effluence and effluence harbours disease as certainly as does any other risk factor in the new environment.

Cancers tend to appear more frequently in post-middle-aged people. Populations of the more industrialized world, including migrants, have longer life spans than those who remain in their places of birth, although even in these places conditions of life and life span are slowly improving as modern sanitation, medical care and lifestyles creep or stride toward full mimicry of the wonders of modernity. Because of longer lives, better access to medical treatment and better statistical accounting of diseases in many nations and ethnic groups, differences between the prevalence of various cancers in sedente, migrant and the established populations in the new locales are being documented. In an Israel-based migrant study, Modan (1981) shows that among migrants there the frequency of types of cancers found is intermediate between that of the place of origin and the host country. He also notes that over time there is a tendency for an increase in cancer at sites where the disease is related to high socioeconomic standards (colorectal, breast, ovarian, gall bladder and uterine corpus) and a decrease in sites related to lower socioeconomic standards: stomach and cervix.

Modan supports his thesis of an association between socioeconomic status and the locus of cancer sites by showing that those sites common among the European-born migrants in Israel are also most common in United States Whites; while those more frequently found among United States Blacks are found relatively more frequently in the non-European migrants to Israel, where, he holds, it is not ethnicity but socioeconomic status which differentiates these populations.

Since others do not specifically note the association of the same cancers with socioeconomic status, Modan's observations must be seen as limited for the present to the populations he has studied. For example, among Asian Indians in South Africa, observed between 1949 and 1969, the overall cancer risk did not change, and remained low (Walker, 1981). Among Asian Indian women in South Africa, as with White women, there was a decline in cancers of the stomach and uterus, while cancer of the cervix and oesophagus rose in frequency. The occurrence of cancers of the oesophagus and stomach were reduced for the Asian Indian males in South Africa, and cancer of the colon, common among the White population, had a crude mortality rate among South African Asian Indians of only 20% of that of Whites.

MacMahon & Pugh (1970) seek an environmental explanation for the differences in cancer rates between those observed among Japanese sedentes, Japanese migrants to the United States and the rates recorded among natives of the host country. They note that in Japan rates for stomach cancer are over five times those in the United States, a frequency which is virtually reversed in the incidence of intestinal cancer. Cervical cancer is over twice as common in Japan as in the United States, while breast and prostatic cancers are relatively infrequent in Japan, but common in the United States. Japanese in the United States, whether born in Japan or of second or subsequent generations born in the United States, show rates of cancer of the cervix comparable to those of United States Whites and different from the Japanese sedentes. The rates for stomach and intestinal cancers among United States Japanese are intermediate between those for Japanese sedentes and those for United States Whites. Despite these shifts in rates, the Japanese-American rate for breast cancer remains more similar to that of the sedentes (relatively rare) than to that of United States Whites (relatively common).

In a study of Chinese, Japanese, Filipino and United States migrants to Hawaii, Glober & Stemmerman (1981) found that both increases and decreases in rates are more marked among men than women and that tumours common in the country of origin decrease in frequency among all

Table 8.6. *Cancer rates per 100 000 people (ages 15–75+)*

Cancer site	Highest rate population	Rate per 100 000	Lowest rate population	Rate per 100 000
Stomach	Male Japanese in Japan	85	Filipino-Hawaiian male	9
	Female Japanese in Japan	40	Chinese-Hawaiian male	9
			Chinese-Hawaiian female	6
Breast	US Caucasians on mainland and Hawaii	80		
Prostate	US Caucasians			
	Mainland	45		
	Hawaii	42		
Colorectal	Caucasian-Hawaiian females	35	Japanese-Japanese females	10
	Chinese-Hawaiian males	49	Filipino-Hawaiian females	15
Nasopharyngeal	US-Chinese males	19		
	Singapore-Chinese females	7		
Lung	Hawaiian-Hawaiian males	71		
	US-Chinese females	22		

(After Waterhouse *et al.*, 1976.)

Table 8.7. *Rates of all cancers per 100 000*

	Hawaii	Mainland United States	Japan	Singapore
Caucasians	291	301		
Hawaiians	279			
Japanese	235		190	
Chinese	234	257		225
Filipinos	143			

(After Waterhouse *et al.*, 1976.)

migrants to Hawaii. These rates continue to fall in succeeding generations, thus supporting the influence of environment on expression. In
Hawaii, as on the mainland, the rate for stomach cancer among Japanese
migrants decreases, as do the rates for nasopharyngeal cancers among
Chinese. These downward trends continue in succeeding generations. At
the same time, those tumours common in the host country tend to
increase in frequency among the migrants to Hawaii, so cancers of the
lung, colon, rectum, breast and prostate are more common among the
Japanese migrants to Hawaii than they are among the Japanese sedentes.

Changes in diet have been suggested as a possible causal factor in the
reduction in stomach cancer (Glober & Stemmerman, 1981), but one
wonders why the same diet would reduce the rate of stomach cancer while
increasing the rate of colorectal cancer.

Although not specifically presented in a socioeconomic framework,
these findings corroborate Modan's observations that some cancer rates
are closely associated with family income. In Hawaii the three groups with
the highest family incomes (Chinese, Japanese and Caucasian) have the
highest male colorectal cancer rates, while the lowest rates for these
cancers are found in the groups with the lowest family incomes (Hawaiian-
born Hawaiian) (Glober & Stemmerman, 1981). On a world-wide basis
there are population patterns for proclivity to develop particular cancers,
but these are apparently neither genetic nor immutable as the rates in
those ethnic groups which have been studied change after migration (see
Tables 8.6 and 8.7).

Migration and mental disease

Psychological maladaptation to new situations is frequently experienced
by migrants. Such inabilities to cope with strange cultures and societies

are not limited to the less affluent segments of the migrant population. Whenever sociocultural pressures become too oppressive, a psychological malfunction may be expected. In some instances migrants bring with them learned appropriate ways of behaviour which seem sufficiently strange to those in the host country to be considered psychologically abnormal (e.g. the North American's maintenance of a continually cheerful countenance in Eastern European countries where smiles are only supposed to accompany humour; the 'forced' laughter of Japanese in Europe or America creates a similar tension). Sometimes such behaviours do not manifest themselves until a life crisis occurs for which no acceptable reaction has yet been internalized: corporate wives unable to cope with Third World expectations when accompanying their husbands on overseas assignments; anthropologists faced with personally insurmountable problems in the field; academics moving to new jobs in unfamiliar settings – such circumstances provide the environment for severe 'culture shock' which, in extreme situations, leads to a 'nervous breakdown'. Furthermore cultural definitions of mental derangement can lead to actions which, in the West, would be regarded as totally abnormal but which might be understandable within the framework, for example, of a Plains Amerind Vision Quest, or the native experience of Arctic Hysteria or running amok. In the Soviet Union the authorities have defined overt criticism of the state as 'insanity'; in the West such behaviour is defined as political expression.

Unfamiliar experiences for the migrant may precipitate stress reactions which leave the newcomer unable to interact satisfactorily in the new situation. The understood norms of the native culture no longer hold and a number of impediments to comfortable participation in the new culture (including language differences, inadequate housing, lack of ties to family or other known associates, etc.) lead to alienation, depression, and emotional and social isolation. Psychobiological responses to these stresses are not uncommon. Much scientific literature notes a relation between stress and hypertension, stress and obesity, stress and cardiac disease, etc., although, as has been noted, it remains a difficult task to quantify stress directly.

In a review of migration and mental health, J. Humphreys (personal communication) notes that in Australian studies the patterns and extent of mental illness are reported as different among migrants coming from varying originating locales. Jews from displaced-persons camps who were sent to Australia experienced great mental difficulties, while migrants from Great Britain or those who were internal migrants within Australia had many fewer such problems (Krupinski, 1984). Those whose new experiences included coping with a new language, a new diet, a new climate, and new moral values were more likely to show signs of mental illness than were migrants from the United Kingdom where there were more points of similarity between the two cultures.

Hertz (1982) notes four risk factors important in triggering episodes of mental illness among migrants:

(1) the extent of estrangement from one's own culture;
(2) the length of absence from one's original home environment;
(3) the migrant's age: the elderly migrant and the adolescent are particularly at risk;
(4) non-marital status: the single, widowed, separated and divorced are most at risk.

Gender is also a risk factor for mental illness, particularly as migrant women have fewer opportunities to become integrated into the new sociocultural setting and are equally unable to re-create in the new locale the interaction patterns which gave meaning to their lives in the home country. Several studies (Abu-Lughod, 1980; Cochrane & Stopes-Roe, 1981; Dean *et al.*, 1981; Krupinski, 1984) note that women migrants who work are paid less, have fewer skills and, in addition to their paid jobs, also have responsibilities for child and home care. Such women migrants are often at the lowest end of the socioeconomic level, and have very restricted social contacts.

One group of migrants whose mental health has been extensively studied has been the holocaust survivors who, as displaced persons bereft of home, family and all tangible ties to their past, found haven in Australia, the United States, Scandinavia, Great Britain and Israel. Early in the years following World War II the holocaust survivors suffered from tuberculosis, gastrointestinal disorders and a variety of psychological problems. Besides neurological and mental illness, premature ageing and frequent illness was common (Eitinger, 1971). In an extensive, longitudinal, medical study of 225 Norwegian concentration camp survivors who returned to Norway, Eitinger found many characteristics which, together, made up what have been referred to as a concentration camp syndrome and which served to describe the survivors in his study. These characteristics were not here complicated by a migrant effect, but they were consistent with findings among survivors who were relocated in Israel, the United States and Germany. The concentration camp syndrome included: failing memory and difficulties in concentration; nervousness, irritability and restlessness; fatigue associated with sleep disturbances; emotional instability; headaches and vertigo; loss of initiative, moodiness and feelings of insufficiency. The emotional and mental disorders observed were found to be a result of anxiety which not only produced psychic trauma, but were also associated with psychosomatic diseases: diarrhoea, peptic ulcer and impotence (T. Thompson, personal communication). Krystal (1971), in a study of 2000 concentration camp survivors living in Detroit, noted many common physical symptoms including the inability to relax, palpitations, hyperventilation, dizziness, fainting and heaviness in the chest. In his study these manifestations of psychological malfunction were associated with the increased incidence of peptic ulcer, diarrhoea and dermatosis. Krystal concluded that survivors age earlier and die younger. These findings hold regardless of where the survivors migrated. Although this population was probably under more stress than many other migrant groups, their post-migrant experiences provide the best evidence available for the stress-related emotional costs associated with forced migration, and with the non-psychological manifestations of stress-based diseases including many which have been considered in previously discussed studies as related to modernization, improved living conditions and more ample diets.

Insect-borne diseases and migration, with particular reference to malaria

Malaria continues to spread in migrant populations where the disease has not yet been brought under control. Even where there have been large-scale eradication programmes, problems of disease control remain in nomadic (i.e. seasonally migrating) populations. In the post World War II years, and particularly through the 1950s, there was great hope for the complete eradication of malaria through the destruction of the breeding sites of the mosquitoes and the use of residual DDT sprays in settlements throughout the malarial regions of the world. Unfortunately, although the World Health Organization (WHO) knew where the malarial areas of the world were, they failed to note the migratory habits of the human hosts (Prothero, 1977). The residual insecticides were very effective in eliminating mosquitoes and other insects from residences for periods up to six months, and those sedentary populations who would agree to additional chemical cures would have had their malaria brought under control. But migratory populations (pastoralists, fishermen, cultivators, migratory labourers, traders, pilgrims, refugees, etc.) were not always where the spray team was at any given time. In many instances, residences were also movable and so could not be sprayed. On the north coast of Peru, in the Province of Lambayeque, people locked their houses and went away for the scheduled spray day in preference to having the teams come and 'mess everything up' with their smelly spray that left a white residue (not to be removed) over everything. The local population had known malaria as endemic to the region and considered the DDT spray treatment as inconvenient rather than providing protection from a life-threatening illness. In North Africa where Prothero served for many years as a consultant to WHO, he noted (Prothero, 1977) considerable migration which the WHO operatives found too complicated to deal with as pastoralists crossed national boundaries in north-western Nigeria. It was very difficult to maintain contact (to continue the eradication programme) even with those initially treated.

Although DDT has been declared a health hazard to humans and its use in anti-malaria campaigns has stopped, the need to pay attention to many forms of internal and external, short term and long term, long distance and nearby migration of population has been clarified through the inadvertent failures of the anti-malarial campaign. Different patterns of mobility have different effects on exposure to disease, on the transmission of disease, and on the effectiveness of programmes to improve public health (Prothero, 1977).

In Africa south of the Sahara the annual return of migrants from their various work places brings many of them through malarial areas, providing a consequent reintroduction into their home communities of fresh malarial infestation, thus maintaining a reservoir of hosts from whom the disease is constantly transmitted (LeRiche *et al.*, 1971).

In west Thailand, near Bangkok, the annual influx of migratory agricultural labourers has also been associated with reinfection with malaria (Sormani *et al.*, 1983). Migrants were more at risk than were the sedentes in the villages studied, especially those whose cane-cutting assignments took them to wetter areas. Only the poorest of the sedentes went periodically to malarial areas to make charcoal or to dig for ore. Up to 44.7% of those who went on these journeys had malaria. The authors imply that the migrants know the disease risk involved in labour migration as they state that in this rural tropical area economic constraints force people to expose themselves to the vector of this disease.

Malaria is not the only disease that extends its range as the human host disperses. For example, Prothero (1977) says that cholera spread from west to east across Africa, beginning in the Republic of Guinea, where the disease was introduced in August 1970 by an infected student returning by air from the Soviet Union, and in 1975 the movement of Bangladeshi refugees led to the spread of smallpox in that country and into the eastern areas of India.

240   B. A. Kaplan

Asthma in migrants

Several studies document an increase in the prevalence of asthma among migrants. Tokulauan migrants to New Zealand experience an increase in asthma (I. Prior, personal communication), and children of Italians settling in Sydney, Australia, have higher asthma rates than do natives of Sydney (Peat *et al.*, 1980). In Sydney more asthma is found in middle-class families with two or three children than in larger families or poorer households. In a study in London in 1977 I found that a startling 30% of Asian migrants interviewed reported development of asthma following their migration. Another study in the north of England (Smith *et al.*, 1977) also documented higher asthma rates among migrants. A few asthmatics who migrated to Florida from within the United States improved, but more either were not improved or were made worse (Chen, 1968). Some South Africans who moved to coastal cities from inland found their breathing difficulties exacerbated after visiting or living on the coast, and reported either improvement or cessation of symptoms on their return inland (Ordman, 1956, 1959). Immigrants to Israel from Iran experienced breathing difficulties shortly after arrival in their new setting, though those from other countries did not have the same reactions (Gutman, 1958). And in Japan, following World War II, substantial numbers of American military forces in the Tokyo–Yokohama area had their first asthmatic attack after being in that region for a short time (Huber *et al.*, 1954; Phelps *et al.*, 1961; Smith *et al.*, 1964). A large number of migrants to Taiwan from mainland China also had their first asthma attack in the new setting (Chen, 1968). In this Taiwanese study more migrants reported experiencing their first attack in middle age, while the sedente Taiwanese with asthma were more frequently young adults. Of those with seasonal allergies and asthmas, 80.1% were migrants, whereas non-seasonal asthma was common (62.4%) among the sedentes. Six years following migration, however, there was a shift in the distribution of asthma problems with a change from predominantly seasonal allergies and asthmas to a gradual approach to the more typical non-seasonal distribution of the Taiwan sedentes. It would appear that the asthmatics gradually adapted to the new environment and climate. This study also provides support for the hypothesis that, as regards their illness and disease proclivities, migrants bring with them the adaptations that were appropriate in their home environment, but, with time, come to resemble more closely the sedentes. Finding this pattern repeated in so many settings and for so many diseases provides a strong case for adaptation as an important factor in the diffusion of people and diseases, once more showing how malleable a creature is *Homo sapiens*.

Migrant's perception of their health status

In situations where it is not possible to collect as much anthropometric or laboratory-based information as one might like, the responses of the subjects themselves often yield satisfactory information about their condition, their methods of treatment, and the outcome. In fact, Baker (1981) reports that respondent's replies are often more accurate and specific in detailing health conditions, as well as emotional reactions, than physicians' reports or other more 'objective' quantifiable results. In one such study Baker & Beall (1982) collected the responses of high-altitude Peruvian migrants living at low altitude through the use of a questionnaire and compared these results to similar responses from high-altitude sedentes, migrants from one low-altitude community to another and to sedentes at the new locale. All respondents were asked whether they had been ill in the previous two weeks; if so, with what, and what treatment they had sought. Those exposed to greater environmental change were more affected. High-altitude migrants to low altitude reported significantly more gastrointestinal symptoms than others. These disturbances are associated with their labour in irrigated agricultural fields at low altitude, where the environment is more conducive to the spread of diarrhoeal diseases than are the dry high-altitude zones to which they were accustomed.

Nearly all causes of death are reported more frequently among high-altitude migrants to low altitude and they have an overall death rate nearly twice that of low-altitude natives. Vehicular deaths and poisoning (often with unfamiliar agricultural chemicals) are principal causes of accidental deaths at low altitudes. Furthermore, death from respiratory disease is three times greater for high-altitude natives than for low-altitude natives, but high-altitude natives have substantially lower rates of cardiovascular diseases than do low-altitude natives.

In another questionnaire-based survey of over 19000 individuals in 3900 households of migrants and sedentes in Teheran, it was found that males of all ages reported similar sickness rates without regard to their place of birth: sedentes (17.0%) and migrants (18.2%) both reported having been sick during the previous two weeks (Stromberg *et al.*, 1974). Female sedentes matched the male rates (17.6%), but female migrants had a significantly higher rate of reported sickness (26.9%). In the same study, adult women also reported more psychosomatic symptoms and/or psychiatric problems than did adult males. Migrants of lower social status had higher sickness rates than others and used medical facilities less, but more skilled migrants who had higher social status were more likely to use the medical facilities. The relationship between social status and the use of medical facilities does not hold for the sedentes (Stromberg *et al.*, 1974). Migration to Teheran has been a long tradition in Iran involving traders, pilgrims, those moving in from unproductive lands and those who are still participating in nomadic patterns which bring them into Teheran from time to time. The greater difficulty that women find adjusting to a more restricted lifestyle in Teheran has already been noted.

Hispanic migrant farmworkers were studied in Wisconsin to determine their medical needs and their utilization of medical facilities (Slesinger & Cautley, 1981). The farmworkers were asked to evaluate their state of health, to indicate which of a list of possible medical conditions they had experienced, and what sort of medical resources they had used. Of the population, 90% had Spanish as their primary language and had little education, and the great majority had very low incomes. Only 17% of the group considered themselves in excellent health, compared to 32% of the low-income population in the nation. Older Spanish monolingual women with less education considered their health to be worse than that of others. And, as has been reported in other studies, the women migrant workers consistently regarded themselves as afflicted with troublesome conditions to a greater extent than the males. In the self-evaluation of ailments by all migrants, the highest ranked problems were (in order): headache, eye trouble, backache, tooth or gum problems, nervousness, irritability, difficulty sleeping, coughing, stomach pain and low spirits. (It should be noted that none of the 24 medical conditions to which they were asked to respond were traditionally used Hispanic health categories.) It is clear, however, that psychological problems and difficulties associated with lack of access (or perceived lack of access) to medical care ranked high. In fact, only 57% had received medical care in the year before the interviews, well below the 76% of low-income families who got such attention in the national sample. A larger proportion of women than men had seen a physician, as had more older than younger workers. These visits had been for general physical examination, orthopaedic and musculoskeletal problems, minor illnesses and infections, skin disorders and gastrointestinal and digestive problems. There were few consultations for diabetes, cardiovascular and respiratory problems (Slesinger & Cautler, 1981).

Discussion

It would appear that individual physiological adaptation is established early in life in the environment in which one lives. This adaptation is not always suitable for the demands of a new environment, but offspring of the migrant population appear to adapt to the locale in which they find themselves, not only culturally but also physiologically. This would explain why those born in one geographic locale have incidence rates different from those migrating there (MacMahon & Pugh, 1970).

The question remains whether the diseases of old age are apt to increase everywhere as the life span is increased, or only as modernization effects brought on by changes in lifestyle, work habits and nutritional preferences become more widespread. Or will the migration effects thus far reported as affecting the most recent migrants in highest degree tend to taper off in subsequent generations, perhaps as those emotional stresses which exacerbate overeating, hypertension and associated diseases are reduced as, for example, somewhat reduced rates of CHD are observed in established modern populations today?

Conclusions

From a diverse range of studies using several study models, all of which had in common the examination of migrants' health (or lack of it) it is clear that not only do migrants and all travellers bring with them their ills from their homeland, some of which spread to the new locales, but they also – in time – come to adapt to the illness gradients which are more typical of the places to which they move. Some of the dramatic increases in prevalence rates which have been documented for migrants may, therefore, be expected to diminish in time as the migrants become more acclimated to their new environments and as medical knowledge concerning the treatment of these ills becomes better able to deal with problems that are not amenable to 'simple' adaptation.

Acknowledgment. I should like to thank the Department of Physical Anthropology of the University of Cambridge and Churchill College, Cambridge, for providing facilities which made possible the completion of this chapter. I should also like to thank the editors, Gabriel W. Lasker and C. G. N. Mascie-Taylor, for their advice and assistance. I am responsible for any errors that may remain.

References

Abu-Lughod, J. (1980). Migrant adjustment to city life: the Egyptian case. *Urban Life: Readings in Urban Anthropology*, ed. G. Gmelch & W. Zenner, pp. 77–89. New York: St Martin's Press.
Baker, P. T. (1981). Modernization, migration and health: a methodological puzzle with examples from the Samoans. Paper presented at the Conference on the Consequences of Migration, East-West Center, Honolulu, Hawaii, August 25–31, 1981. 27 pp. mimeo + 13 tables and figures.

244   B. A. Kaplan

Baker, P. T. & Beall, C. M. (1982). The biology and health of Andean migrants: A case study in South Coastal Peru. *Mountain Research and Development*, 2, 81–95.

Chen, Cheng-yen (1968). Studies on bronchial asthma in migrants from the China Mainland to Taiwan. *Memoirs of the College of Medicine of the National Taiwan University*, 13, 17–36.

Cochrane, R. & Stopes-Roe, M. (1981). Psychological symptom levels in Indian immigrants to England – A comparison with the native English. *Psychological Medicine*, 11, 319–27.

Cruz-Coke, R., Donoso, H. & Barrera, R. (1973). Genetic ecology of hypertension. *Clinical Science and Molecular Medicine*, 45, supplement 1, 55s–65s.

Cruz-Coke, R., Etcheverry, R. & Nagel, R. (1964). The influence of migration on blood pressure of Easter Islanders. *Lancet*, 1, 697–9.

Dean, G., Walsh, D., Downing, H. & Shelley, E. (1981). First admission of native born immigrants to psychiatric hospitals in Southeast England, 1976. *British Journal of Psychiatry*, 139, 506–12.

Eitinger, L. (1971). Organic and psychosomatic after effects of concentration camp imprisonment.' *International Psychiatry Clinica*, 8, 205–17.

Gerber, L. (1984). Diabetes mellitus among Chinese migrants to New York. *Human Biology*, 56, 449–57.

Glober, G. & Stemmerman, G. (1981). Hawaii ethnic groups. In: *Western Diseases: Their Emergence and Prevention*, ed. H. C. Trowell & D. R. Burkitt, pp. 319–33. Cambridge, Mass.: Harvard University Press.

Gutman, M. J. (1958). An investigation into environmental influences in bronchial asthma. *Annals of Allergy*, 16, 536.

Hertz, D. G. (1982). Psychosomatic and psycho-social implications of environmental changes on migrants (a review of contemporary theories and findings). *Israeli Journal of Psychiatry and Related Sciences*, 19, 329–38.

Hoobler, S. W., Tejade, C., Guzman, M. & Pardo, A. (1965). Influence of nutrition and acculturation on the blood pressure levels and changes with age in the Highland Guatemalan Indian. *Circulation*, 32, supplement II, 116 (abstract).

Hopkins, D. R. & Florez, D. (1977). Pinta, yaws and venereal syphilis in Colombia, *International Journal of Epidemiology*, 6, 349–55.

Huber, T. E. *et al.* (1954). New environmental respiratory disease (Yokohama Asthma). Preliminary Report. *American Medical Association Archives of Industrial Hygiene*, 10, 399–408.

Krupinski, J. (1984). Changing patterns of migration to Australia and their influence on the health of migrants. *Social Science and Medicine*, 18, 927–37.

Krystal, H. (1971). Trauma: Considerations of its intensity and chronicity. *International Psychiatry Clinics*, 8, 11–28.

LeRiche, W. H., Harding, W. & Milner, J. (1971). Time and place. In *Epidemiology as Medical Ecology*, ed. W. H. LeRiche, W. Harding & J. Milner, pp. 259–62, Edinburgh: Churchill Livingstone.

Livingstone, F. B. (1958). Anthropological implications of sickle cell gene distribution in West Africa. *American Anthropologist*, 60, 533–62.

MacMahon, B. & Pugh, T. F. (1970). *Epidemiology: Principles and Methods*. pp. 175–85. Boston, Mass.: Little, Brown.

Modan, B. (1981). Israeli migrants. In *Western Diseases: Their Emergence and*

Prevention, ed. H. C. Trowell & D. R. Burkitt, pp. 268–84. Cambridge, Mass.: Harvard University Prees.

Nadim, A., Amini, H. & Malek-Afzali, H. (1978). Blood pressure and rural–urban migration in Iran. *International Journal of Epidemiology*, 7, 131–38.

Neel, J. (1962). Diabetes mellitus: A thrifty genotype rendered detrimental by progress? *American Journal of Human Genetics*, 14, 353–62.

Ordman, D. (1956). The climate factor in perennial respiratory allergy and its relation to house dust sensitivity. *International Archives of Allergy*, 9, 129–45.

Ordman, D. (1959). The climate group of respiratory allergy patients. *International Archives of Allergy*, 12, 162–9.

Peat, J. K., Woodcock, A. J., Leeder, S. R. & Blackburn, C. R. (1980). Asthma and bronchitis in Sydney. II. The effect of social factors and smoking on prevalence. *American Journal of Epidemiology*, 111, 728–35.

Phelps, H. W., Sobel, G. W. & Fisher, N. E. (1961). Air pollution asthma among military personnel in Japan. Some of the clinical characteristics of this disease and therapeutic measures that seem to give patients most relief. *Journal of the American Medical Association*, 175, 990.

Prior, I. & Tasman-Jones, C. (1981). New Zealand Maori and Pacific Polynesians. In: *Western Diseases: Their Emergence and Prevention*, ed. H. C. Trowell & D. P. Burkitt, pp. 227–67. Cambridge, Mass.: Harvard University Press.

Prothero, R. M. (1977). Disease and mobility: A neglected factor in epidemiology. *International Journal of Epidemiology*, 6, 259–67.

Savitt, T. L. (1978). *Medicine and Slavery: The Diseases and Health Care of Blacks in Antebellum Virginia*. Urbana: University of Illinois Press.

Scotch, N. A. (1963). Sociocultural factors in the epidemiology of Zulu hypertension. *American Journal of Public Health*, 53, 1205–13.

Slesinger, D. P. & Cautley, E. (1981). Medical utilization patterns of Hispanic migrant farmworkers in Wisconsin. *Public Health Reports*, 96, 255–63.

Sloan, N. R. (1963). Ethnic distribution of diabetes mellitus in Hawaii. *Journal of the American Medical Association*, 183, 419–24.

Smith, R. B. W., Kolb, E. J., Phelps, H. W., Weiss, H. A. & Hollinde, A. B. (1964). Yokohama asthma, an area specific air pollution disease. *Archives of Environmental Health*, 8, 805.

Smith, J., Morrison, L., Harding, K. & Cumming, G. (1977). The changing prevalence of asthma in school children. *Clinical Allergy*, 1, 57–61.

Sormani, S., Butraporn, P., Fungladda, W., Okanurak, K. & Dissapongra, S. (1983). Migration and disease problems: A study of patterns of migration in an endemic area of malaria in Thailand. *Southeast Asian Journal of Tropical Medicine and Public Health*, 14, 64–8.

Stromberg, J., Peyman, H. & Dowd, J. E. (1974). Migration and health: Adaptation experiences of Iranian migrants to the City of Teheran. *Social Science and Medicine*, 8, pp. 309–23.

Walker, A. (1981). South African Black, Indian and Coloured populations. In *Western Diseases: Their Emergence and Prevention*. ed. H. C. Trowell, & D. R. Burkett, pp. 285–318. Cambridge, Mass.: Harvard University Press.

Waterhouse, J., Muir, C. S., Correa, P. *et al.* eds (1976). *Cancer Incidence in Five Continents*, vol. 3. Scientific Publications 15. Lyon: International Agency for Research on Cancer.

Glossary

Acculturation. Cultural adaptation by immigrants into a different society or by both groups during fusion of demes. The term is sometimes used of cultural changes through modernization of a whole society.

Cline. A gradual change in gene frequencies with distance over a broad geographical area.

Deme. A subpopulation within which most mating occurs. The smallest endogamous subunit of a population. In human populations the boundaries of *demes* are often poorly demarcated.

Demic diffusion. A combination of gene flow and invasion involving only partial displacement and mixture of an immigrant population with the population already there.

Dendrogram. A graphic branching model of population origins based on the assumption that fission largely accounts for their genetic relationships to each other.

Deterministic model. A model which assumes a population of infinite size and thus no random elements.

Displaced person. A refugee of World War II Nazi concentration camp or other person so designated by United Nations provisions for refugees from war.

Distance model. Conceptualization of migration as being systematically influenced by geographic distance from place of origin to destination.

Environmental growth factor. The cause of the additional growth in stature, length of long bones and some other body dimensions caused by long-term improvements in standard of living such as ample available food.

Fission. The split-up of a village into two 'daughter' demes which are genetically isolated from each other by the migration of one or both.

Fusion. Aggregation of two or more demes with summation of their gene pools.

Gene flow. The process by which genes gradually spread from group to group by local exchange of mates in more or less continuously inhabited space.

Heterosis. The concept that offspring of genetically different parents grow more vigorously and are bigger.

Host population. A population or deme, typically a dominant one, receiving immigrants.

Invasion. Expansion of a group from its original range into unoccupied territory or by displacement of another population.

Island model. Conceptualization of migration as taking place between pairs of well-demarcated demes. The matrix of migration between demes.

Isolation by distance. In the absence of geographic barriers, the differentiation of a population (especially in respect to gene frequencies) by dearth of gene flow.

Kin-structured migration. Movement of families or other related individuals, resulting in the non-random contribution of genes of the donor gene pool to the recipient population.

Labour migration. Migration, typically of single males, for a limited time as temporary periodic unskilled workers.

Marital migration. The movement measured by the distance from place of residence of one spouse to that of the other at the time of their marriage.

Migrant worker. Labour migrant, especially agricultural stoop labourer who follows the harvests from one climatic zone to the next. Women and children as well as men are involved.

Plasticity. The ability of the organism to respond adaptively to changes in the environment. The term is usually confined to developmental changes in childhood that produce long-lasting or permanent improvements in the capacity of the organism to cope.

Refugees. Individuals forced to emigrate by economic, political or military conditions in their native country.

Returnees. Migrants who left a population but later returned.

Rural-to-urban migration. Migration from smaller places to cities involving increased outbreeding and radical changes in lifestyle.

Sedente. A person remaining behind in the original location after fissioning or steady emigration of part of a population.

Selective migration. In the neo-Darwinian sense, the emigration of individuals who are not a genetically random representation of the gene pool of the population.

Stochastic model. A model in which the finite size of the population is taken into account and thus random elements play a part.

Stress. A term used with several meanings. Environmental stress is the impact of milieu, for instance as experienced by migrants, which tends to unsettle physiological functions.

Stress is also used in a psychological sense to describe a feeling of unease and instability. Such meanings overlap, since psychic stress may be accompanied by such physiological responses as increased catecholamine excretion rates.

Index

Aborigines (*see also* Australia *and* New Guinea), Australian, 36, 138, 140; New Guinea, 36
abortions, spontaneous, 223
acclimatization, 189; altitude, 184
acculturation, 177, 198, 246; acculturated status, 194; *in situ* acculturation, 197; to Western society, 114
activity: load changes, 223; patterns, 95, 113, 118
adaptation, 48, 108, 120–1, 167–8, 172, 175, 180, 191, 194, 199, 203–4, 216, 220, 222–3, 240; adaptive strategies to urban life, 107, 121; biocultural patterns of, 183; biological, 93, 167, 204; biosocial, 95; coastal, 24; cold-adaptation studies, 176; developmental, 167, 176; short-term, 177; failure in, 181; genetic, 177, 184, 189, 219, 228; individual physiological, 242; marine economic, 32; migrants' lack of, 167; new modes of, 168; of fertility to urban migration, 48; phenotypic, 95; short-term, 167; 'simple', 243; state of, 168; theory, 167; thermal, 176; to high altitude, 183, 189, 191; to local diets and activity patterns, 95; to past environments, 94; to the city, 118; to the diseases of urbanization, 110; urban, 112, 120
admixture, 134, 144, 148, 151, 158–9
aerobic capacity, 199
Africa and Africans (*see also* West Africa), 15, 23–4, 36, 48, 52, 99, 105, 108, 116, 131, 135–7, 141, 144, 149, 151, 153, 155, 158–9, 178, 180, 224, 239; African migrant groups, 177; East, 216; extreme north-east, 136; non-Bantu populations of, 149, 153–4; North, 216–17, 224, 226, 239; south of the Sahara, 239
Afro-Americans, 177; low socio-economic level, 178
Afro-Caribbeans, 180
agriculture, 23, 131, 223, 232; agricultural populations, 143; agricultural workers,

111; agricultural states, 148, 150, 151; indigenous agricultural peoples, 150; swidden agriculturists, 140–1
Ainu of Hokkaido, Japan, 23, 144, 146, 150
Alaska, 17, 24–8, 30, 32, 35, 137; Bristol Bay, 25; interior, 15; south-western, 25, tundra areas of, 147; Umnak Island, 25
Aleuts, 17, 24, 27–34, 140; Aleut-Eskimo (*see also* Eskimo), 30–2, 35, complex, 36; linguistic unity, 30; population system, 31
alleles, 20, 54, 70, 142; allele frequencies, 20; allelic genes, 71
altitude, 10, 112, 171, 187; acclimatization, 184; changes in, 222; effects, 170; high-altitude, 171, 185, 191, 241; 'intermediate', 171; low-altitude, 171, 185, 187, 241; medium-altitude natives, 185
Amazon, 138; Amazonia, 143
America, 14, 16, 18, 20, 25, 32, 34, 46, 50, 57, 93, 116, 137, 143, 145, 148, 151, 219, 236; American aboriginals, 145; American children, 172–3; American colonies, 45; American genes, 151; American gene pool, 149; American migrant children in Rio, 173; American Revolution, 44; Latin, 99, 105–6, 108, 151, 229; native plants in, 23; prehistory of, 25; tropical, 15
Americans, 14; earliest, 15–17, 29; early prehistory of the, 25; native, 29
American Indian (Amerindian), 14, 16–17, 20, 22, 27, 29, 32–5, 52, 146, 149, 153, 155, 202; Amerindian genes, 151; Amerind population, 219; indigenous, 53; North American Indian tribes, 30; North American Indians, 31, 220; Pima Indians, 114; unmigrated, 35; Yanomama, 140–1, 146; Yanomama villages, genetic relationships among, 141
Amish, 43, 47; in Holmes County, Ohio, 46–7; Old Order Amish, 46

248

Anadyr River (*see also* Siberia), 26–7, 30
analogy, 133–4
ancestors: ancestor frequencies, 71;
    common ancestors, 79, 131, 157;
    migrant, 58
ancestry: 'Caucasian', 150; common, 79,
    81–3, 86, 142; European, 190;
    Hawaiian-born subjects of European
    and Japanese, 177; human racial
    ancestral history, 156; shared, 81, 87
Andes, 107, 138, 182, 184; Andean
    migration, 182; Central, 182
anthropometric data, 133
Antilles, Lesser, 52
Apache, 25, 147
Arawak, 52
archaeology, 29
Arctic, 18–19, 34, 136, 140, 147; Arctic
    Circle, 16; Arctic Hysteria, 236
Argentina, 105
Arizona, 148
Arnhem Land, 36
Asia and Asians (*see also* South-east Asia),
    14, 24–5, 91, 99, 105, 108, 138, 147, 150,
    153–4, 159, 224, 226–227, 229, 232,
    240; eastern, 32, 35–6, 150;
    immigrations from, 145; migrants to the
    United Kingdom, 181; northern, 32, 34;
    north-eastern, 14; South Asia, 180;
    South Asian children, 180–2; South
    Asian migrants, 203; South Asian
    populations, 32, 180; south-eastern, 14,
    23, 33; south-east continental, 136, 138
assimilation: to United States urban
    culture, 115
asthma, 197, 203, 240; asthmatics, 240;
    childhood, 197; non-seasonal, 240;
    problems, 240; rates, 240; seasonal,
    240
Athabascan Indians (Athapascans), 28–31,
    145–8; ancestors, 147; Athapascan
    salient, 148; California and British
    Colombia, 148; dialect groups, 27–28;
    language, 25, 30
atherosclerosis, 223, 226, 231
Australasia, 138; Australasians, 153
Australia and Australians, 5, 14, 16, 18–19,
    20–4, 32–3, 35–6, 42, 58, 131, 138, 150–1,
    159, 216, 222, 236–7, 240; Australian
    Aborigines, 20, 22–3, 33; central, 20;
    coastal colonization of, 24; northern,
    20, 36; populations, 21; south-eastern,
    23
Australoids, 14
Azerbaijani villagers, 111
Aztec: areas, 145

Bandkeramik, 144
Bantu, 155; Africans, 149; agricultural,
    149; Bantu source population, 155;
    Bantu West Africa, 153
basal metabolic rate (BMR), 11, 175–6
Basques, 144
Bering-Chukchi platform, 26
Bering land bridge (platform), 15–17, 25,
    30–1, 36, 137, 146
Bering Strait, 25, 36
Bering Sea, 16–17, 24, 26, 32; Mongoloid
    population on, 17, 27–9
Beringia, 145, 147; southern Beringian
    coast, 147
bifurcation (*see also* divergence), 27–8, 31,
    157; separation/bifurcation model,
    160
biology, 167; biological adaptation, 93;
    biological adaptive patterns, 167;
    biological fitness, 198, 203; loss of, 168;
    of South Pacific migrants, 203; biological
    problems, 94; biological selection (*see*
    selection); biological taxonomy, 134;
    human, 94, 118
biome: mammoth-steppe or arctic-steppe,
    26
birthplaces of husband and wife, 74–6
birth rate, 74, 105–6
Blacks (*see also* Afro-Americans), 152,
    224; in New York City, 226–7; in urban
    South Africa, 230; middle-aged, 226;
    population of the United States, 153;
    population in South Africa, 226–7, 229
Black Caribs, 52–3
blood groups, 3, 6, 8, 11, 35, 152;
    geographic distribution of, 6–9;
    blood-group studies, 14; Diego blood
    group, 52; Duffy blood group, 53; Gm
    blood-group system, 53; MNS
    blood-group system, 52
blood pressure (*see also* hypertension), 11,
    110–14, 119, 177–80, 186, 197–9, 202–4,
    224–6; blood-pressure elevation, 210,
    228; levels, 111; rates, 224; rise after
    migration, 113; diastolic, 197;
    difference, 112; low or normotensive,
    204; of urban migrants, 113; sedente,
    114; systolic, 178, 180, 197–9, 224;
    systolic and diastolic, 177–9, 186, 194,
    199, 201
body: build, 99, 177, 225; composition, 97,
    175–6; dimensions, 246; length
    measurements of body segments, 11;
    mass (obesity), 230; size, 102, 117, 192;
    temperatures, 175
body weight, 113, 174, 176, 178, 194–5,

body weight—*continued*
198–200, 202, 223–4, 229; at birth, 190;
differences, 195
Bolivia, 183–5, 189; Bolivian research, 190
Bombay, 176
Britain (Great Britain), 7–8, 42, 47, 58,
216, 219, 237; eighteenth-century, 101;
migrants from, 236
British Honduras, 52
British Isles, 149
Brownian movement model of migration, 5
Bushman (*see also* San), 141, 144

calcium (*see also* metabolism, calcium),
181–2
California, 173, 176, 198, 200–1, 228–9
Canada, 27, 46, 145, 151; immigrants into,
45; north-eastern, 31; north-west part of,
148
cancer, 116, 119, 232–5; breast, 233–5;
cervical, 233; colorectal, 234–5;
intestinal, 233; lung, 220, 234–5;
nasopharyngeal, 234–5; prostatic, 233–5;
rates, 223, 235; differences in, 233; sites,
232–3; stomach, 233–5
cardiovascular disease (disorders) (CVD),
191, 194–5, 197, 199, 225, 229, 241–2;
cause of death from, 186; CVD rates in
Samoa, 202
catecholamines, 179, 180; excretion of, 12,
179; excretion rates, 179, 247; in
migrants, 180; urinary, excretion rates
of, 116, 180
Caucasians (*see also* Whites), 150, 153,
235; Caucasoid, 152–3, 155; in Hawaii,
228
Central American, 23, 52, 151, 216
cephalic index, 4, 11
change (*see also* culture change *and*
environmental change), 168; activity
load, 223; age changes in blood pressure,
177, 198; biological and behavioural,
198; biomedical, 222; diet, 202, 223; in
lifestyle, 199; social, lifestyle and
environmental, 183; spatial, 168; time,
168
chapattis, 180–2
chest size, 191
childbirth, 182
children, 100, 173; activity patterns for,
118; African immigrant, 180; American
(United States), 172–3; American-born,
of European immigrants, 172;
California, 173; child survival, 107;
Chinese, 173; Chinese immigrant, 180;
European, 173; European urban, 181;

from Glasgow, 180; Guatemalan, 101,
173; Japanese, 173; Korean, 173;
London Chinese, 173; migrant, 99, 101,
112, 187; high-altitude migrant, 189;
Pakistani immigrant, 182; rural, 99; rural
Indian, 100; school, blood pressure of,
112; sedente, 187; size of, 168; South
Asian, 180–1; Westernized, 173; Zaire
Hutu, 174
Chile, 110, 184, 189–90, 225
China, 115–16, 220, 240; north, 33, 36;
south, 115
Chinese, 115, 150, 172, 176–7, 229, 235;
born in New York, 115; born in the
United States, 115; children, 173;
China-born Chinese, 115;
Hawaiian-born, 231; living in the United
States, 115; males in Hawaii, 228;
migrants to Hawaii, 231, 233; migrants
to New York City, 115, 231; migrants to
the United States, 115; southern, 22
cholecalciferol, 180
cholesterol: levels, 112
Choukoutien: Lower Cave of, 34; 'Old
Man' of, 35; Upper Cave of, 24, 34
city, 91, 93–6, 99, 101, 118, 204; adaptation
to the, 118; characteristics of the, 94;
inner city, 91; lifestyle of the, 117;
long-term residence in the, 120;
mediaeval, 91; New World, 92; of Third
World nations, 94; Old World, 92;
'optimal city size', 96; outer city, 91;
stress of the, 121; Western, 197
climate, 147, 158–9, 221, 240; effects of,
on north-south genetic clines, 142;
North American, 217, 218
clines (gradient sequences), 6, 8, 28, 36,
144, 156, 246; clinal maps, 131, 156–7;
cranial, 31; north-south genetic, 142;
of blood groups, 9
cold (*see also* adaptation), 107, 176, 182;
cold-adaptation studies, 176; cold stress,
192
Colombia, 105
colonization, 149, 182; European, 151;
long-distance, 148
comparative anatomy, 132–3
concentration camp: Norwegian,
survivors, 237; survivors living in
Detroit, 238; syndrome, 237
consanguinity, 58; consanguineous union,
55
continuity, 132, 137; dental, 159; genetic,
62, 156; morphological, 159; of hominid
species, 160; 'racial', 159

contraception, 48; contraceptive technique, 105
Cook Islands, 113–14; Cook Islands Maoris, 194
Cornwall, 72, 74–6, 79, 81–2; migration in, 43
coronary heart disease (CHD) (*see also* heart disease), 109, 111, 113, 115–16, 119, 217, 220, 223, 226–8, 231; coronary risk factors, 201, 203; mortality rates from CHD in Hawaii, 228; rate, 111, 228, 243
Costa Rica, 105
Creoles, 52
cultivators, 238; Hutu, 175; swidden, 140–1; traditional, 171; Turkana, 179
culture, 43; advanced, pre-state, agricultural stage of human, 143; change, 168–9, 179, 204; cultural data, 133; cultural differentiation, 144; cultural values, 105, 111; 'culture shock', 236; Neolithic, 143; pre-state, 143
Cumbria, 64–5, 87

Dakar, 112
death, 167, 185–6, 191, 226, 231, 241; causes of, 185–6, 241; death rate, 186, 226, 241; age-adjusted CVD, 201; American Samoa CVD, 201–2; from stroke, 224, 226–7
demes, 130, 140–1, 246; adjacent, 137; 'daughter' demes, 246; demic expansion, 136, 140; discrete, non-overlapping demes, 70; local, 136, 141–2; parental, 136; temporary, 142
demic budding, 136–7
demic diffusion, 6, 130, 144–5, 147–9, 152, 159, 246; large-scale, 143; model, 144; systematic, 160
demic structure, fluid, 140, 141
demography, 43, 103; demographic-transition theory, 47; of immigrant communities, 46; of migration, 204
dendrogram, 20–1, 81, 131, 139, 145–6, 153, 155–7, 246; construction, techniques of, 152; of relationship, 153
Denmark, 48–9
dental: gradients, 150; morphology, 36; variation, 24
dentitions, 24, 159
deoxyribonucleic acid (DNA), 2, 4; genomic DNAs, 152; mitochondrial, 152–3, 155; nuclear, 155
descent, 131
deterministic model, 79

development, 117, 119, 121, 168, 172; biological change in, 103; economic, 178; human (physical), 95, 97–8; motor, 187; of state societies, 148; rural, social and economic, 96; Western European industrial, 168
Devon, 74–6, 79, 81–2
diabetes, 115, 119, 194, 196–7, 199, 201, 203, 217, 223, 228–231, 242; diabetes mellitus (Type 2 diabetes), 12, 109, 113–15, 202–3; genetic susceptibility to, 202; insulin-dependent, 229; mortality (death) from, 115, 231; non-insulin-dependent, 228–9; prevalence of, in the Pacific, 202
diet, 110, 113, 118–19, 173, 175, 180, 202, 222–3, 231, 235, 237–8; carbohydrate, 225; dietary changes, 217, 223; energy, 195; experience, 230; fat, 195; problems, 223; protein content, 226; status of high-altitude migrants and sedentes, 187; improved, 11, 99; migrant, 187; practices in Iran, 111; patterns, 115; South Pacific Islanders' traditional, 202; traditional South-Asian, 182; dietary intake, 191; dietary intake, high-altitude, 186
differentiation: human, 32; linguistic, 29; of a population, 246
discontinuity, 133, 148
disease (*see also* cardiovascular disease *and* respiratory diseases), 46, 91–3, 95, 109–111, 116, 172, 185–7, 192, 194, 216–21, 223, 228–9, 231–2, 239–40; childhood, 219; communicable, 218–19; deficiency, 180; degenerative, of middle age, 229; disease control, 238; epidemic, 105, 219; hereditary, 219; human, of mixed aetiology, 3; immunities to, 95; industrial, 220–1; infectious, 107, 109–10, 117, 119, 218; insect-borne, 238; kidney, 224; minor diseases, 115, 119; non-local, contagious diseases, 220; old age, 243; of Westernization, 223; organisms, 96; parasitic, 117; patterns, 168, 221; proclivities, 240; quarantinable, 219; rates, 222; relationship between migration and, 219; resistance, 20; risks, 222–3; risks of migration, 223; states, 180; stress-related, 238; transmission, 216, 220, 239; vectors of, 10, 216; Western, 220, 223
displaced persons, 237, 246; camps, 236
displacement, 1, 144, 147, 152, 246; demic, 140; partial, 246

disruption, 108
distance, 131, 246; models, 70, 246
divergence (*see also* bifurcation), 27–9;
Aleut/Eskimo, 28; Athabascan/Bering
Sea Mongoloid, 28; genetic, 134;
linguistic, 25, 29; niche, 147; population,
25, 29; times, 155
diversity: genetic, 28, 157; human, 130,
134, 160; patterns of, 160
diverticulosis, 231–2
DNA, *see* deoxyribonucleic acid
donor country, 10; drift, 71, 83, 155; drift
effects, 83

Easter Island, 110; Easter Islanders, 20;
Easter Island migrants to Chile, 225
ecosphere, human, 4
ecosystem: grazing, 26; rural, in Iran, 111;
urban, in Iran, 111
Ecuador, 185
education, 104, 106, 108, 111, 116, 119,
225, 242; women's, 106
emigrants, 2, 43, 47; returned, 10
emigration, 1, 44, 48–51, 53–5, 57, 74, 192,
247; effects of, on the gene pool, 53;
from Europe, 57; from the Irish
countryside, 92; net, 48; of families, 59;
rural, 117; selective, 2, 117
endogamy, 75, 77–8, 83, 87; island, 77, 83;
level, 58; rates, 79; same-island, 83
energy, 195; dietary, 195; intake, 186;
nuclear, 221; storage, 192
England, 42–4, 102, 109, 116–18, 217;
north of, 240
English, 151
environment, 2, 4, 10–12, 103, 107, 115,
167–8, 172, 175, 204, 216, 219–22, 232,
236, 240–2, 247; American urban, 98;
biological and sociocultural, 222;
changed, 10–11; changes in the, 198;
changing, 168; high-altitude, 171;
Hutu-migrant, 175; malarial, 3; new,
167, 169, 180, 243; new work, 220;
non-malarial, 3; pairs of, 11;
pre-migration, 102; rural, 118;
temperature extremes, 175; unhealthy,
223; urban, 94–9, 104, 108, 116, 120,
169; Westernized, of urban Teheran,
111; Westernized, sociocultural, 229
environmental changes (*see also* changes),
12, 98, 168, 172, 181, 183, 198–9, 203;
effects of rapid, 192, 194
environmental conditions, 11
environmental risks, 221, 223
environmental stress (*see also* stress),
167–8, 172, 247

environmental variables, 2–3; urban, 119;
environmental variability, 10
equilibrium, 83; genetic, 61; situation, 66;
values, 79, 83, 87
Eskimo, 16–18, 24, 27–35, 140–1, 146–7;
Alaskan, 176; Aleut-Eskimo stock (*see
also* Aleuts), 30–1, 35; linguistic unity,
30; population system, 30; Bering Sea
coast Yupik, 32; Eskimo-Aleut, 145–6;
Eskimo-Aleut separation, 146;
Eskimo-Amerindian cluster, 31;
Greenland Eskimo, 24; language, 30;
last Eskimo migration into Greenland,
18; Polar, 29
Ethiopia, 217
ethnicity, 171, 187, 190, 223; South Asian,
181
Eurasia, 131, 136–7, 156; Eurasians, 154
Euro-Americans, 178
Europe, 24, 44, 58, 91, 104, 143–4, 149,
169, 217, 236; central, 159; Eastern and
Central, 144; Eastern European
countries, 236; European genetic map,
149; European expatriates, 176;
industrial cities of Western, 181;
northern, 149, 153; north-west, 144;
urban areas of, 110; Western, 159;
Western European industrial
development, 168
Europeans, 35, 52, 148, 150–1, 154, 176,
220; ancestors, 53; ancestry, 190;
colonization, 151; expatriate, 176;
north-west, 180; in South Africa, 227
evolution, 94; evolutionary agenda, 15;
evolutionary speciation, 134;
evolutionary theory, 167; human, 15, 32;
molecular, 155
exchange: exchange rates, 78; genetic, 71;
inter-island, 83; marital, 71, 78, 82, 86;
mate exchange, 71–2; of individuals, 70
exogamy, 61, 75; local band, 137; rates,
79; spatial, 61
expansion, 130–1, 136, 138, 148–50, 246;
demic, 136, 140; of Western disease,
223; population, 130, 158; territorial, 219

family: emigration of, 59; income, 235;
newly migrant, 187; nuclear, 63; size,
106–8; unstable family relations, 223
famine, 217–18
fat intake, 197
fatness, 113–14, 119, 229
fauna: Australian Pleistocene, 16; land, 17
fertility, 47–8, 50–1, 93, 95–6, 103–8,
117–19, 121, 185, 191; age cross-over in,
105; differentials, 105; effects of

high-altitude hypoxia on, 185; fertility differences from general population, 47; fertility effects of modest emigration, 49; goals, 108; high migrant ferility, 105; historical and political influences on, in Morocco, 106; intermediate, 105; longitudinal studies of fertility adjustment, 107; of migrant women, 105, 107; of rural sedentes, 105–7; of urban migrants, 105–6; of urban natives, 105–6; patterns, 47; adjustment to host-fertility patterns, 47; preference, 48; relationship of, to family movements, 42; rural, 108; surveys of women in Nuñoa, 185; total population fertility, 48; trends in India, 47
fertility rate, 48, 105, 107; age-specific, 47; of Korea, 108; total, 108; urban, 104
Fiji, 114
Filipino, *see* Philippines
fission (fissioning), 59, 141, 246–7; demic, 140; fission/fusion process, 141; group, 141; initial, 60; lineal fissions, 60; local, 141; 'tree' of fission products, 141
food, 18–19, 96, 117, 182, 186–7, 221, 231; additives, 232; food-consumption habits, 187; imported, 203; junk, 231; new, 113; preferences, 4; products, 187; shortages of, 11, 101; sources, 136; storage and distribution services, 96; supply, 92; surplus, 168; vitamin-D-fortified, Western, 182
food-sharing, 16
foodstuffs, 10
Fore, 22
fossils: Asian fossil specimens, 137; evidence, 138; fossil human remains, 32; fossil man, 24; record, 132
founder effect, 192
France, 109, 112, 151, 176, 202, 216–17; southern, 16, 58
French Pyrenees, 58
fusion (fusioning), 59, 141, 246; demic, 140; fission/fusion process, 141

Gambians, 178
gastro-intestinal diseases (disorders, disturbances), 231–2, 237; gastro-intestinal problems, 242; symptoms, 241; system, malfunction of, 231
gatherers, 94
genes, 2, 6, 14, 31–2, 36, 51–5, 57–62, 64, 67, 70–2, 130, 133, 140, 142–3, 148–9, 151, 160, 246; allelic genes, 71; Amerindian, 52, 151; geographical distribution of, 11; indigenous, 144; local, 149; loss or fixation of, 61; marker, 20, 23; movement of, 130; of aboriginals, 148, 151; of immigrant origin, 58; random migration of, 2; recombination of, 6; single-gene analysis, 32; substitution, 72
gene flow, 2, 6, 36, 61–2, 64, 66, 70, 130, 133–5, 137, 140, 145, 152, 156–8, 160, 246; channelled, 147; intercontinental, 156; inter-demic, 134, 140; intertribal, 149; local, 131; models, 13
gene frequencies, 2, 6–7, 11, 27, 30, 36, 52–3, 57–8, 61–3, 71–2, 130–2, 139, 141–2, 144, 157, 246; affinities, 30; clines, 130, 142, 144; data, 133, 143, 152, 160; differences between Scotland and the rest of Britain, 7; east–west clinal gene-frequency pattern, 156; gradients of, 140, 150; multilocus gene-frequency gradient, 144; patterns, 131, 142, 144, 156; pattern of multi-regional or multi-tribal, 141; picture, 156; smooth clines of, 142; variation, spatial, 71
gene identity, 52
gene movement, latitudinal, 66
gene pool, 2, 4, 6, 52, 54–5, 57–60, 64, 67, 130, 144, 151, 246–7; African, 53; American, 149; equilibrium gene pool, 72; homogenization of the human, 149; of Arthez d'Asson, 58; Latin American, 151
General Household Survey, 42
generations, 60, 80–1, 84, 86–7, 112, 243
genetic: adaptability (*see also* adaptation), 53; anomalies, 221; barriers, 131, 143; composition, 70; constitution, 53–5; 68; defects, 223; differences, 12, 52, 59, 140, 144; differentiation, 11, 20, 30, 31, 61, 135, 140–1, 147, 152; distance, 20–2, 30, 72, 139, 146, 154; drift, 2, 63, 72, 130, 134, 140, 192; effects (*see* immigration, genetic effects of, migration, genetic effects of, *and* movement, genetic effect of); endowment, 2; exchange, 71, 75; factors, 115, 177; heterogeneity, 68; homogeneity, 2, 62; identity, 43; maps, 148, 149, 152; markers (*see also* genes), 35, 143, 155; 'noise', 141; pattern, 155; plasticity, 169; prehistory, 27, 29; separation, 21; structure, 55, 57, 61, 62, 68; theory (population), 60; time-clock, 28–9; variability, 52, 60, 67, 68, 143; genetics, 51, 171; multivariate, 158; population, 6, 14, 25, 61
genotypes, 2, 11, 54; European, 144;

genotypes—*continued*
haemoglobin genotypes AS and AC, 53;
European, 144
geographical homogeneity, 7
geographical location, 7, 10
Germany, 44, 216–17, 237
Glasgow, 180
glucose tolerance, 230
gout, 109, 194
gravitation model of migration, 5
Greenland, 18, 27, 30–1; Greenland
Eskimo, 24
growth, 95, 97–101; 104, 117–19, 121, 172,
175, 187, 189; adolescent, 173, 185, 187;
biological change in, 103; changes, 99;
child, 100, 185, 187, 190; delays, 100;
differences, 102–3; environmental
growth factor, 11, 246; growth data, 173;
infant, 187; intrinsic, 136–7, 140, 148;
intrinsic local, 138; longitudinal, 173; of
Hawaiian-born Japanese, 98; of Italian
migrants, 97; of Japan-born Japanese,
98; of Japanese children, 98; of Japanese
sedentes, 98; of rural-born Germans in
Baden, 97; of United States children,
172; patterns, 190; perinatal and
postnatal, 189; physical, 119; postnatal
growth studies, 190; pre-adolescent, 173;
secular trend of, 173; urban, 105
Guatemala, 52, 100, 120; Guatemala City,
99–100, 107, 121, 173; Guatemalan
children, 101, 173; rural-urban migrants
in, 225

haematocrit, 112
haptoglobin pattern, 138
haplotype, 52, 154; African, 154;
distribution, 155; Eurasian, 154;
pattern, 154
Hawaii, 98, 101, 113–14, 116, 192,
199–201, 216, 222, 228–9, 231, 233, 235;
Chinese migrants to, 231, 233; Filipino
migrants to 180, 231; growth of Japanese
children in, 98; Hawaiian Islands,
227–8; Hawaiian-born subjects of
European and Japanese ancestry, 177;
Hawaiian Japanese, 228; Japanese males
in, 228; migrants to, 235; rural,
traditional areas of, 114; total cash
economy of, 114
Hawaiians, 228, 235
head: dimensions, 172; shape, 4, 11, 169
health, 95–7, 109–10, 113–14, 116–17, 168,
172, 185–7, 192, 194, 198, 203, 216–18,
222, 241–3; care, 99, 104; changes, 116;
characteristics, 118; conditions, 241;

impairment, 199; maternal, 189; mental
health, 236–7; problems, 221; status,
109, 113, 116
heart disease (*see also* coronary heart
disease), 115, 226; atherosclerotic,
226; cardiac, 227, 236; ischaemic, 226–7
heterogeneity, 71
heterosis, 98, 246
heterozygosity, 53
Himalayas, 171
Hippocrates, 109
HLA: antigens, 22; Australoid HLA
profile, 22; gene-frequency distributions,
22; genetic complex, 22; HLA-All.B40
linkage relationships, 22; system, 22, 144
Holland, 144
Holy Island, Northumberland, 63
hominids, 33, 131, 135–6, 155; hominid
population, 221; hominid species,
competing, 135; hominid species,
continuity of, 160; opposition, 138
homogeneity, 24, 31, 36, 71–2, 79–82, 85–6
homozygosity, 61; homozygote FyFy, 53
Hong Kong: age-mates, 173; Chinese, 173
Honolulu, 114, 198
host: countries, 217, 233, 236; host
community, disease risks characteristic
of, 222; reproductive norms of, 48; host
factors, 20; host population (*see also*
population), 52, 67, 70, 167, 204, 246;
human, migratory habits of the, 238
humidity, 10
Hungarians, 43
hunters, 17, 26, 94; game, 137; Mesolithic,
144; migratory abilities of early, 16; of
Australia, 23
hunter-gatherers, 42, 134, 136–7, 140–1,
143, 147, 155, 159; levels, 151; local,
population, 143; nomadic, 167
hunting, 15–16; hunting habits, 18
Hvar, Dalmatian island of, 45, 59
hybridity, 83; coefficient of, 72, 83, 85;
hybridization, 159
hygiene, 175
hypertension (*see also* blood pressure),
109–12, 119, 177–8, 191, 194, 196–7, 203,
217, 223–5, 229, 231, 236, 243
hyperurisaemia, 194, 203
hypoxia, 107, 182, 189–91; high-altitude,
185

illness, 167–8
immigrants, 2, 36, 44–6, 67, 98–9, 147, 182,
240, 246; African, 53; African and
Chinese immigrant children, 180;
'changes in the bodily form of

descendants of', 98; European,
American-born children of, 172;
immigrant community, 47, 67;
immigrant communities, demography of,
46; populations, 53; male immigrant
group in Cornwall, 43; status, 181; to
Western cities, 182; transatlantic, 45
immigration, 1, 52, 58, 79, 86, 138, 147,
192; contested, 130; effects of, on the
recipient population, 117; from Asia,
145; genetic effects of, 52; into New
Hampshire, 44; into South Carolina, 44;
long-range, 59, 66, 83; quotas, 218
inbreeding: derived, 63; pedigree, 88
inbreeding coefficient: autosomal, 55–6;
sex-linked, 55–6
Incas: Inca areas, 145; Inca bone, 33;
Inca times, 182
income, 106, 110
India, 36, 47, 114, 116, 150, 176, 180, 216,
224, 230, 239; central, 150; diabetes in,
230; fertility trends in, 47; Indian
immigration into the United Kingdom,
47; Indian subcontinent, 219; natives of,
176
Indians (*see also* Asian Indians, American
Indians, North America *and* South
America), 150, 154, 224
indigenes, 148, 159
Indochina, 176
Indonesia, 18, 35; Indonesian populations
of south-eastern Asia, 14
Indonesians, 159
industry, 221; industrial diseases, 220–1;
industrial societies, 231
industrialization (*see also* society), 223;
industrialized countries, 229;
industrialized world, 223, 232
infants: Indian, 190; Mestizo, 190
infant mortality, *see* mortality, infant
in-migration, 92–3; high rates of, 94;
rural, 91–2; urban, 93
instability, 168, 247; emotional, 237
intermixture, 52
International Biological Programme, 97,
192
invasion, 130, 142, 145, 147–9, 152, 156,
158–9, 246; force in the New World,
149; models, 131, 133, 135;
unchallenged, 135
Iran, 111, 224, 240, 242; Iranian women,
225
Ireland, 43–4, 46, 117–18, 217; western, 87
Irian Jaya, 20
island model, 62, 70, 246
Isles of Scilly, 72–6, 83, 85–8; exchange

rates in the, 78; inter-island exchange,
83, 86; inter-island marriage, 78; island
endogamy, 83; marital exchanges, 82;
population sizes, 77–8, 82; total
population of the, 79
isolates, 17, 61; Anaktuvuk Eskimo, 31;
differentiated, 62; religious, 88
isolation, 30, 62, 74, 77–9, 83, 86–8, 131,
134, 156–7; by distance, 6, 131, 137, 156,
246; cultural, 147; emotional and social,
236; genetic, 58, 134; geographic, 36;
population structure,
isolation-by-distance models of, 62
isonymy: methods, 88; studies, 83
Israel, 102, 224, 226, 232, 237, 240;
European migrants in, 233;
non-European migrants to, 233
Italy, 97; Italians settling in Sydney,
children of, 240; Italian Alps, remote
hamlets in, 88

Jamaica, 178; Jamaicans, 178
Japan, 98, 144, 153, 173, 226, 228, 233, 240
Japanese, 22, 146, 150, 176, 226, 228, 235,
236; children, growth of, 98, 173;
Hawaiian-born, 98; immigrants to
Hawaii, 101, 228, 231; migrants to
California, 176; migrants to the United
States, 233; native-born and
American-born, 172; sedentes, 98, 101,
228, 231, 233, 235; United States
Japanese, 233
Java, 18, 36; 'Java Man', 36
Jews, 3, 42, 149, 236; Jewish population of
New York City, 228

Kenya, 178; Kenyans of the Luo tribe, 113
kin, 136, 145; kin and non-kin networks,
119; kin(-based), 170; kin-structured
migration, 59, 60
kinship, 63, 66, 72, 83–4, 87, 95;
coefficients, 57, 66, 71–2; levels, 87;
values, 83, 87
Karkar, *see* New Guinea
Korea, 48, 173; Korean children, 173;
Korean World Fertility Survey, 48, 108;
second-generation migrant Korean
children, 173
kuru, 22

labour, 241; labour-force participation,
104, 108; Labour Force Survey, 42;
labour migrants, 232, 247; labour
migration, 216, 239, 247; migrant labour,
216; wage, 114
labourers: migratory, 238; migratory
agricultural, 239

Latin America, *see* America, Latin
liberalization, political, 104
life expectancy, 92, 225, 227; at birth, 226;
    in England and Wales, 109; in France,
    109; in London and Manchester, 109;
    in Norway, 109; in Sweden, 109
lifestyle, 111, 113, 171, 180, 202–3, 217,
    231–2, 242–3, 247; attributes, 204;
    changed, 221; changes (*see also* change),
    111–12, 183, 199; stress of, 180;
    concomitants, 115; New York, 231; of
    the city, 117; rural, 9; traditional native,
    194; traditional pastoral, 171; urban, 99,
    118; variables, 115; Western, 115, 178–9
lineage boundaries, 59
linguistics, 132; linguistic data, 133;
    linguistic divergence, 29
living conditions, 110, 181, 238; for the
    migrant Hutu in Katanga, 175;
    unhygienic, 101
locale: migrant donor, 10; recipient, 10
London, 92–3, 95, 109, 179, 240; London
    Chinese children, 173
longevity, 117
Los Angeles, 95, 173
Louisiana, 43
lung function, 190–1
Lyon, 112

Magellanic Islands, 19
Maine, 44
maladaptation, psychological, 235
malaria, 110, 112, 218–20, 238–9; malarial
    areas, 239; infestation, 53, 239; species,
    53; reinfection with, 239
Malaysia, 60
malnutrition, 109–10, 117, 187
mammals, 136; land, 145;
    mammal-hunting, 147; savannah, 137
marital exchange (*see also* exchange), 71,
    78, 82, 86; marital-exchange matrices, 74
marital migration, *see* migration, marital
marital movement, *see* movement, marital
marriage (*see also* migration, premarital,
    marital *and* postmarital, *and* heterosis),
    age at, 104–7; endogamous, 75, 77;
    marriage–distance analysis, 62;
    marriage distances, 64; distribution of,
    61–2; marriage records, 74; residence
    after, 74–5
Maryland, 44
Massachusetts, 44–5
mate exchange (*see also* exchange), 130,
    133–4, 137, 142, 246; demic, 137
maturation, 97, 102, 104, 112, 118–19, 168,
    172

medical services (facilities), 116, 119,
    241–2, medical care, 175, 232, 242; lack
    of access to, 242; medical conditions,
    242; needs, 242
megafauna, 16, 26
Melanesia, 138, 191; Island Melanesia, 21,
    139; Island Melanesian population, 20;
    populations, 21–2; rural, 229
Melanesians, 15, 20, 35, 114, 138; Island,
    22; Papuo-Melanesians, 23
Melanesian-speaking groups, 20
mental illness (disease, derangement), 217,
    235–8
Mesolithic, 159; genes, 144; hunters, 144;
    inhabitants, 144; peoples, 144
Mestizos, 10, 182, 187; infants, 190
metabolism, 175; calcium, 181–2; energy,
    175
Mexicans, 10, 98, 152, 172, 229; sedentes,
    103
Mexico, 10, 98–9; albumin Mexico, 52;
    migrants from, to the United States, 103;
    returned emigrants in, 10; Valley of, 151
Mexico City, 93; migrants from Oaxaca to,
    170
Michigan, 112
microdifferentiation, 140; local, 141;
    tribal, 140
Micronesia, 114, 229; population, 22
Micronesians: Nauru people, 114
migrant (*see also* non-migrants *and*
    rural-to-urban migrants), 1–5, 10, 12,
    16, 18, 26, 30, 42–3, 46–7, 51, 53, 59, 61,
    63, 67, 72, 74–5, 87, 96, 98–103, 105–114,
    116–17, 120–1, 167, 169–73, 176–7,
    179–80, 186–7, 189–91, 201–4, 218–26,
    228–32, 235–7, 239–43, 247; adaptability,
    120; Andean Indian, 204; biology of
    long-term, 120; biosocial status of, 120;
    diets, 187; downward, 171, 189, 191;
    Quechua and Aymara, 189; Indian,
    infants, 190; eastern lowland,
    settlements, 190; European, 170;
    female, 241; Filipino, living on Oahu,
    Hawaii, 116; first-generation, 179; first-
    and second-generation in New York
    City, 169; from Easter Island to Chile,
    110; from Oaxaca to Mexico City, 170;
    from Samoa to Hawaii, 192, 200; from
    South Asia to the United Kingdom, 180;
    from the Philippines to Hawaii, 180;
    from Tokelau Island to New Zealand,
    194; high-altitude, 184–5, 187, 189, 191,
    225, 241; high-altitude, to sea level, 189;
    Hispanic, farmworkers, 242; internal,
    108, 111, 116; international, 1, 116;

Japanese migrants to California, 176; labour migrants, 232, 247; local, 1; long-distance, 191; long-term, 120; low-altitude, 184–5, 191, 226; low-altitude, to high altitude, 189; Mexican, 169; migrant children, 187, 189; migrant diseases, 232; migrant effect, 237; migrant Hutu in Katanga, 175; migrant labour, 216; migrant women, 237; migrant workers, 220; national (internal), 1, 236; natural increase among, 105; offspring of, 221; peoples, 203; permanent, 221; Puerto Rican, 121; recent, 120; rural, 1; seasonal, 41; second-generation, 172–3; (pre-)selection of, 10, 48, 102, 107, 169, 172; South Asian, 203; status, 109; to low altitudes, 188; upward, 184, 189, 191; upward and downward, 170; urban, 1, 97–9, 103, 105, 109, 111–15, 118, 121, 225; social selection of, 117
migrant groups, 238; African, 177; high-altitude, 186, 191; low-altitude, 186
migrant populations (*see also* populations), 32, 47, 169, 175, 192, 221–2, 230, 236, 238, 242; migrant-population studies, 193; migrant Samoan populations, 199; occult morbidity in, 197; South African, 230
migrant studies, 221, 223, 231; Israel-based, 232; migrant-sedente study, 224; migrant-study design, 175, 221; Polynesian, 179
migration (*see also* rural-to-urban migration), 1–8, 10–12, 18, 25, 31, 36, 42–5, 47–8, 51, 53, 57, 59–61, 63, 67, 71, 75, 79, 83, 86–7, 91, 94–5, 99–103, 108–10, 112–21, 130–2, 134, 136, 139, 143, 157, 160, 167–70, 172–3, 175, 177, 182, 192, 194, 199, 202–4, 216–18, 220–3, 225, 228–9, 231, 235, 238–40, 242, 246; age at, 100–4, 112, 171, 189; Andean, 182; animal, 16; biology of, 67; biological aspects of, 1; biological consequences of recent, 120; biological effects of, 5, 117, 167–8; channelled migration effect, 138; demography of, 204; direct, 105; disease risks of, 223; downward, 183; effect of, on rate of differentiation, 62; effects of, 117, 243; environmental effects of, 3; external, 239; flow, 107–7; forced, 91, 238; forced mass, 1; from high altitude, 186; genetic effects of, 3, 6, 57–9, 68; human, 2, 4–6, 10, 16, 18, 23, 58, 149, 183, 217; models of, 70; hypothetical, 5; in Europe, 104;

in the less developed countries, 104; in the Philippines, 106; individual, 6; intercounty, 7, 9; internal, 7, 42, 239; international, 1, 93; into the Americas, 16; into the New World, 16; into the real Arctic, 34; intrapopulation, 61; kin-structured, 60, 68; labour, 1, 216, 247; lateral, 183; 'laws of', 97, local, 61; long-distance (range), 6, 57, 59, 68, 72, 151, 239; long-term, 239; marital, 2, 6, 71, 79, 247; mass, 131; model, 172; north of Peking, 34; north-south migration track, 147; postmarital, 2, 6; premarital, 2, 6; random, 1, 2; relationship between, and disease, 219; repeated (recurrent), 5, 20, 42; reverse, 111; selective, 1–2, 100–1, 103, 112, 169–70, 247; selective cyclical, 48; short-distance (range), 2, 6, 57–9, 68; short-term, 239; single, 20, 25; sites, 232; social and economic literature on, 117; stress of, 111, 186; to low altitudes, 225; to the United States, 115; trauma of, 118; unstructured, 60; urban, 108, 113, 118, 225; voluntary, 91, 116
migration data, 66; non-directional, 67
migration density: function, 59
migration distance, 64–6
migration matrix, 71, 75–6, 84, 87
migration patterns, 64, 72, 87
migration rates, 70, 103
migration research, 103; Latin American, 120
migration studies, 1–3, 5, 10–11, 91, 97, 169, 171, 204, 222
mobility, 1, 168, 219, 239
modernization, 104, 106–7, 113–14, 118, 168–9, 177–8, 192, 198–9, 203–4, 219, 223, 228–9, 238, 243, 246; diseases of, 229, 231; process, 198
Mongoloids, 14, 35, 138, 150, 152–3, 155; Asiatic, 34; Bering Sea, 17; modern, 33–4, 36; peoples (populations) of eastern Asia, 31, 35; proto-Mongoloid, 35
Mongols, 149; Manchurian, 176; prehistoric, 35
Moors, 149
morbidity, 116, 121, 197, 199; new patterns of, associated with Westernization, 192; occult, in migrant populations, 197; urban, 113
morbidity rates, 110; for Whites and Blacks in the United States, 113
Morocco, 106
morphology, 112, 153; morphological data, 139

mortality, 92–3, 95–6, 109, 113, 117–19,
121, 185, 191, 202; childhood, 109, 185;
CVD-related, 202; from diabetes, 115;
infant, 107; infant, in Sweden, 109;
infant and childhood, 104; infant,
childhood and adult, 109; patterns of,
of low-altitude natives, 185; perinatal,
182; urban, 93, 113
mortality rates, 92, 109–10, 119–20, 186,
227; crude, of cancer of the colon in
South Africa, 233; for Whites and Blacks
in the United States, 113; from CHD in
Hawaii, 228; infant and young child, 190;
urban, 95, 104
movement (*see also* population
movements), 67, 192, 218; between
populations, 70; direction of, 62, 65;
forced, 218; genetic effect of, 64; local,
67; long-distance, 139; marital, 72, 86;
maternal and paternal, 67; movement
patterns, 86; patterns of bodily, 221
mutation, 130, 134–5; rates, 135, 157

Nagoya City, 173
National Child Development Study, 6
National Health Service Central Register,
42
natural increase, 94, 103, 117; among
migrants, 105
natural selection, 5, 98, 130, 134, 140, 192
Nauru, 202
Navaho, 25, 147
Neanderthals, 159
Near East, 144
Negroes (*see also* Afro-Americans and
Blacks), 15; Negroid, 152–3, 155
Neolithic: culture, 143; farmers, 143–4;
peoples, 134, 144; times, 74
Netherlands, The, 226
neuropsychological stimulation, 112
New Britain, 22
New Caledonia, 138
New Guinea, 18–24, 35, 138, 140;
Aborigines, 36; coastal, 20; Highlands,
22, 23; Island of Karkar, 85–7; natives,
150; New Guinea populations, 20
New Hampshire, 44
New Mexico, 10
New World, 14, 16–17, 24, 30, 32–3, 35–36,
42–3, 58, 93, 131, 137, 145–7, 149, 151,
159, 177, 217; analysis, 32; Arctic, 146;
cities, 92; inhabitants, 34; Syndrome,
202
New York, 44, 95, 115, 169–70, 172, 177,
226, 228–9, 231
New Zealand, 47, 113, 192, 194, 196–7,

216, 224, 229, 231, 240; Europeanized
cities in, 113; Maoris, 194–5, 229–30;
migrant population in, 47; Tokelau
migrants to, 113, 192, 195, 198; White
population in, 48, 194
Nicaragua, 52
niche: divergence, 147; specialization, 147
Nigeria: north-western, 239
nomads, 178; native Orochon, 176;
nomadic populations, 216; pastoral
nomadic populations, 178
non-identifiability, 134; problem, 143–4,
160
non-migrants (*see also* sedentes), 1–3, 120,
170, 194, 198, 204; design, 171;
non-migrant natives, 221; rural, 48, 117
non-Whites, living in New York, 115
North America, 17, 36, 140, 145, 147, 151,
202, 216–19; Indian population of, 24;
northern, 19
North Americans: in Eastern Europe,
236; natives, 27; White, 10
Northumberland, 62, 63
Norway, 109
Nuñoa, Peru, 183–7; Nuñoa economy, 184;
Nuñoa research, 185
nutrition, 104, 107, 112, 171, 185–6, 192;
nutritional preferences, 243; nutritional
status, 97, 187

obesity, 12, 109–10, 113–16, 119, 177, 192,
194–5, 197, 199, 202, 217, 223–5, 228–31,
236; chronic, 203; early onset of, 223
occupation, 106, 111, 222, 225, 229;
occupational specialization, 168
Ohio, 46
Old World, 93, 158–60; cities, 92
Omaha Amerindians, 105–6
origin, 27–8, 53
Orkney Islands, 78
Orochons (*see also* nomads), 176
osteomalacia, 180–2
Outer Hebrides, 78, 83, 87

Pacific, 20, 22–4, 34, 131, 139-40, 202, 216;
islanders, 191–2; islands, 138, 192–3;
migrants, 192; migration, 191;
Pacific-island populations, 192, 202;
prevalence of Type 2 diabetes in, 202
Pakistan, 180, 216; Pakistanis, 180, 224
Palaeolithic: Far Eastern Palaeolithic man
of late Magdalenian, 35; Palaeo-Indians,
145, 147; 'Palaeo-Indian' descendants,
145, 148; Palaeo-Indian habitation, 145;
Upper, 16

Panama Canal Zone, 173; Panama-born children, 173
panmixia, 4, 72
Papua New Guinea, 20, 138; Eastern Highlands of, 22
pastoralists, 171, 238–9; East African traditional, 171; Rwanda Tutsi, 175; Samburu, 178; traditional nomadic Turkana, 179
pedigree: inbreeding, 88; studies, 83
Peking, 34; 'Peking Man', 36
Pennsylvania, 44; Meadowcroft rock shelter, 137, 145
personal-achievement orientation, 104
Peru, 10, 183–5, 189, 225–6; high-altitude migrants, 241; high-altitude populations, 223, 225; Indians, 10; north coast of, 238; southern, 183–4
phenotypes, 7, 9, 12, 53; frequencies, 9; phenotypic expression, 167, 169–70
Philippines, 106; Filipinos in Hawaii, 180, 228, 231, 233; natives, 150, 235
phylogeny, 131, 157; models of phylogenetic relationship, 134; phylogenetic history, 155; phylogenetic trees, 133, 153, 157
physical size, *see* body
pig, 23
pigmentation (*see also* skin), 15, 35–6
Pitcairn Island, 46
Pizarro, Francisco, 182
plasticity, 121, 247; genetic, 169; human developmental, 98, 118, 172; of BMR, 176; of human growth, 101
Pleistocene, 16; Australian Pleistocene fauna, 16; late Pleistocene, 19; late Pleistocene migrants into the New World, 17; middle, of Asia, 15
Poland, 99, 102
pollution control, 181
Polynesia, 222, 229; Polynesian diets, 195; groups, 194–5; studies, 202; population, 22
Polynesians, 22, 33, 46, 139, 146, 194, 197; migrants, 112, 229–30; migrants to New Zealand, 232; migrant studies, 179
pongids, 155
population (*see also* host population *and under the names of countries and islands*), 1, 25, 32, 35, 42–3, 49, 53, 60, 94, 131, 151, 154, 167–9, 176, 192, 243, 246–7; aboriginal, 150; advanced, pre-state, agricultural, 143; Alaskan, 29–30; American and Asian, 145–6; Amerind, 219; Asian, 176; Asian (Indian-Pakistani) in South Africa, 224;

Bantu source population, 155; Black Carib, 53; coastal subpopulations, 184; donor, 1–2, 4, 90; expansions/replacements, 158; genetic dynamics of, 61; genetic isolation of human, 134; genetic structure of, 62; high-altitude Peruvian, 223, high-altitude, sedentary, 184; hominid 221; hunter-gatherer, 143; immigrant, 53, 246; Indonesian, of south-eastern Asia, 14; interpopulation variability, 67; intra-population, 67; isolated relict, 144; Latin American, 151; migrant, 32, 169–70, 175, 192, 194, 204, 224, 229, 232; migrant in New Zealand, 47; migrant-population studies, 193; migrant, sedente and returned, 222; migratory, 238; nomadic, 216; non-obese, 202; of origin, 167; originating, 48; Pacific, 202; parental, 67; pastoral nomadic, 178; pressures, 192; recipient, 1–2, 4, 52, 57, 67, 90, 117; rural, 108–9, 117; non-migrating, rural, 109; sedente rural, 225; Samoan, 229; sedentary, 238; sedente, 110–11, 170, 204, 223–5, 229, 232; in north-west Iran, 224; low-altitude sedente, 225; non-Western, 204; settled, 178; size of founding, 20; small-band, 94; Third World, 108; Tokelau Islands, 194; traditional Samoan, 199; tribal, of South-east Asia, 150; urban, 108–9, 117; sedente urban, 225; world's, 169
population density, 19, 92, 143, 151–2
population differentiation, 133
population divergence, 25, 29
population dynamics, 136, 140; local, 135
population expansion, 130, 158
population genetics, 6, 14, 25, 60, 61
population growth, 91, 99; rates, 169; urban, 92, 103
population history, 74, 135
population increases, 183
population movement, 135, 149, 167, 192
population size, 36, 43, 49, 55, 66, 77, 154, 157; effective, 62–3, 83, 135; finite, 71; infinite, 71; of the mediaeval city, 91; reduction in, on Tristan da Cunha, 54
population structure, 32, 62, 70, 167, 219; isolation-by-distance models of, 62
poverty, 105, 117; rural, 110
pre-migrants, 194, 203
primates, African savannah, 135–6
Proust, Marcel, 131–2, 135
public health, 239; facilities, 116; problems, 217; services, 119
Pueblo Indians, 5

260   *Index*

Puerto Rico, 105–6, 120; Puerto Rican
migrants, 121; Puerto Ricans, 172
Pukapuka: Pukapukans, 195, 197–8
Pygmies, 141, 144; of the African
rain-forest, 140

quarantine, 46; regulations, 218
Quechua Indians: Andean, 183; lowland,
176; residents, 189; Quechua and
Aymara, 190
Quest, Plains Amerind Vision, 236

race: human racial ancestral history, 156;
race-separation times, 158; racial
prejudice, 113
raciology, 11
Rarotonga, 195; Rarotongans, 195, 198
receiving country, 10
recipient population, *see* population,
recipient
recombination, 155
records, 41; marriage, 74; parish, 74
refugees, 1, 238, 246–7; Bangladesh, 239;
refugeeism, 1
religion, 43
reproduction, 48–51; effects of hypoxia on,
185
residence, 176, 187, 220, 238; after
marriage, 74–5; altitude of, 185; effects
of change of, 171; high-altitude, 189;
length of, at sea level, 184; length of
urban, 110–12, 116; life-long, at high
altitude, 11; long-term, in the city, 120;
patrilocal, 75; place of, 7; tropical, 173;
urban, 48; urban-rural, 171
residents, 190, 224; at sea level in the
Tambo Valley, 185; long-term, in
Teheran, 111; long-term, in West Africa
and Indo-China, 176; native, 222;
tropical, 176; urban, in Teheran, 225;
urban, in Puerto Rico, 106
resources, 10, 136, 218; medical, 242;
vertical diversity of, 182
respiratory diseases (*see also* asthma), 185,
191, 241; respiratory causes of death,
185; respiratory problems, 242
restriction enzyme: polymorphisms, 154;
polymorphism patterns, 154; site, 154
returnees, 222, 247
rickets, 180–2, 203
Rio de Janeiro, 173
Rome, 92; Roman Empire, 149
rural-to-urban migrants (*see also* migrants,
urban), 48, 93, 104, 107–9, 111, 117,
120–1; adaptability of, 121; biological
selection of, 90; biology of the, 120;
social profile of likely, 118

rural-to-urban migration (*see also*
migration, urban), 48, 90–1, 93–7,
108–9, 116, 119, 121, 178, 247; biological
effects of, 97; effects of, on health, 110;
effects of, on women in Latin America,
106; literature on, 117
Russia, 102, 217
Rwanda, 175; Hutu, 174; Hutu cultivators,
175; Hutu-migrant environment, 175;
migrant Hutu, 175; sedente Hutu, 175;
Tutsi, 174; Tutsi pastoralists, 175

Samoa, 47, 113–14, 191–2, 194; American
Samoa, 114, 201–2, 222, 229;
population, 229; Samoan Islands, 229;
Samoan migrations, 198; traditional
Samoan lifestyles, 202; Western Samoa,
114, 179, 200–1, 222, 229
Samoans, 113, 198–9, 229; American
Hawaiian, 199; American Samoan men,
199; American Samoan migrants,
198–201; American Samoan sedentes,
198, 201; American Samoan women,
114; American Samoans, traditional,
199–200, 202; Californian Samoans, 199,
201; in Hawaii, 229; low levels of
physical fitness of, 199; migrant, 201,
229; modernized, 201; Samoan migrants,
198, 222, 229; Samoan Migrant Study,
198; migrants to Hawaii, 114, 199;
sedentary, 201; sedentes, 114, 198–9,
229; traditional, 201; urban-living, 113;
Western, 198–200
San (Bushmen), 144; of the Kalahari, 140
sanitation, 92–3, 217, 232; facilities, 221;
sanitary practices, 218
Scandinavia, 216, 237
schistosomiasis, 112
Scotland, 43–4, 102
secular change, 10, 168, 173
sedentes, 2, 10, 67, 99–102, 110–11, 117,
170–2, 187, 189, 191, 199, 221–2,
224, 239–42, 247; adults and children in
Europe, 169; children, 187; Eastern
European, 102, female, 241; groups,
173; low-altitude sedente groups, 186;
high-altitude, 184, 187–9, 191, 226, 241,
high-altitude sedente diets, 187;
highland, 187; Japanese, 98;
low-altitude, 184–5, 191, 226; Mexican,
103, 169; non-migrant, 221; populations,
223; populations in north-west Iran, 224;
rural, 96–100, 103, 106, 108–9, 111–12,
114, 117–18, 121, 225; biology of, 117;
fertility of, 105; rural island, fertility of,
106; social profile of likely, 118; status,
109; Taiwan, non-seasonal distribution

of the, 240; Tutsi and Hutu, 175; urban population, sedente, 225; village, 112; women, high-altitude, 185
sedentary behaviour, 115
sedentism, biological correlates of, 118
selection (*see also* natural selection), 11, 71, 102, 107–8, 135, 144, 155, 158, 167, 220; biological, 101–2; differential selection pressures, 4; lack of, or minimal, 172; neutrality to, 11; social, 121, 130
Semai Senoi, 60
Senegal, 111; rural, 112
separation/bifurcation model, 160
separation times, estimated, 155–7, 160; assignments, 157; migrational implications of recent, 160; race (racial), 158
Serer (people in Senegal), 111; migrants to Dakar, 112; pastoralists, 178
serum: cholesterol, 194–5, 228; lipids, 113; serum-lipid values, 210; proteins, 152
settlement, 91, 138, 190; agricultural, 91; density, 148; enclaves, 46; history, 143; initial, 135, 140; nucleated, 168; pattern, 138–9; permanent, 230; settled community, 64, 67
settlers, 138; Australian, 139
Siberia, 16–17, 25, 30, 32, 150; aboriginal populations of northern, 31; Anadyr region of, 25; eastern, 26
Sicilians, 3
sickle cell: carriers, 3; trait, 218, 220
skeleton, 3, 24, 34, 36; skeletal affinities, 159; data, 152; gradients, 150; materials, human, 15; remains, 23, 137
skin, 180; colour, 14, 180–2
skinfold: thickness, 199; triceps skinfold, 199–200; values, 195
slavery, economic, 217
slaves, 57; African, 218; slave-trade period, 177; slave-trading, 218; West African, 178
slums, 99
social class, 102; mobility to higher, 102
social correlates, 120
social mobility, 106
social profile, 118
social selection, *see* selection, social
social services, 104, 116
social status, 241
social stratification, 168
society (*see also* industry *and* industrialization), 246; band, 95; chiefdom, 95; fission and fusion pattern of, 59; state, 148; tribal, 95; Western, 114; Western, diseases of, 217; Western

industrialized, 217, 223; with records, 41
socio-cultural attributes, 204
socio-economic: class, 92; differences, 106, 118; standards, 232–3; status (SES), 92, 99–104, 108–9, 117, 119, 121, 178, 190, 222, 233; low, 113, 181; values, 105; variables, 48
South Africa, 42, 50, 103, 224, 226–7, 233; Asian Indian males in, 233; Asian Indian population in, 226, 230, 233; Asian Indian women in, 233; Black population in, 226, 230; middle-aged Asian Indians in, 230; South Africans, 227, 230, 240; urban, 224; Whites in, 226, 230, 233
South America, 15, 23, 36, 137, 140, 145, 151, 202, 216–17; Central Andes of, 182; Indians in, 187, 190; Indian population of, 24; sedente, high-altitude Indians in, 188; South American lowlands, 140; women of European and Indian ancestry, 190
South Atlantic, 49, 54
South Carolina, 44
South Pacific, 113; Islanders' traditional diet, 202; migrants to urban areas, 113
South-east Asia (*see also* Asia), 150; South-east Asians, 138; tribal populations of, 150
Soviet Union, 236, 239
species: single-species perspective, 32; single-species thesis, 33
stability, 135; in racial characteristics, 11
statistical difference, 133; computation, 132
stature, 3–4, 97, 101–2, 117–19, 169, 172, 174–5, 187–8, 191, 246; intermediate, 175
stochastic models, 71, 83
stress, 95, 116, 119, 121, 168, 171, 179–80, 198–9, 223, 228, 236, 238, 247; biological, 1; cold, 192; constellation of, 169; emotional, 231, 243; environmental, 94, 167, 172, 247; high-altitude, hypoxic, 183; levels, 116; of changed environments, 11; of lifestyle change, 180; of migration, 111, 186; psychic, 247; psychosocial, 177; reactions, 236; stress-related diseases, 223
stroke, *see* death from stroke
subsistence, 114, 171, 178, 194
Sudan, The, 217
Sumer, 91
Sunda, 18–19, 34; shelf area, 19
support services, 96
support systems, 119, 217

262   Index

Sweden, 109; northern, 78
Switzerland, 46, 103, 216; Swiss, 172
Syndrome: Acquired Immune-Deficiency
(AIDS), 220; New World, 202

Tahiti, 46
Taiwan, 240; aboriginals, 150; sedente
Taiwanese, 240
Tambo Valley, Peru, 183–7, 189, 225;
studies, 189
Tanzanians, 178
taro plant, 23
Tasmania, 18–20, 22, 35; Tasmanian
population, 20; Tasmanians, 22; now
extinct, 23
technology, 142–3, 151
Tecumseh, Michigan, longitudinal study,
112
Teheran, 111, 224–5, 241–2
temperature, 10–11; body, 175;
environmental, 175; extremes of, 175;
hand, 176; regulation, 175; water, 18
Teotihuacan, 92
territoriality, 133
Texas, 229
Thailand, 105; west, 239
Third World, 94, 216; expectations, 236;
populations, 108
Thule, 18, 30
Tibetan migrants, 112
Tierra del Fuego, 137
time, 20, 135; changes, 168
Tokelau (Pacific Polynesian Island), 47,
194, 196; Tokelau Island Migrant Study,
192, 194, 222, 224; Tokelau Island
populations, 197; Tokelau Resettlement
Programme, 194
Tokelauans, 47, 194–8; Tokelauan
migrants, 113, 195–8, 229–30, 240;
female, 197; male, 197; Tokelauan
sedentes, 195–6, 229–30; sedente
children, 224
Tolais, 22
Tonga, 191
tools, manufacture of, 16
Torres Straits, 23; Islands, 22
trade, 220; traders, 238, 242
transect, rural, 62
Tristan da Cunha, 49–50, 54, 58–9, 88;
direct reproductive effect of the two
relatively large emigrations, 51;
population, 50; population size, 49;
growth curve of the, 50; Tristan
population, intrinsic rate of natural
increase in the, 51
tuberculosis, 109, 112

Turkana: cultivators, 179; natives from
north-west Kenya, 178; pastoralists, 179
Turkey, 216
typhus, 46

Union of Soviet Socialist Republics
(USSR), 27
United Kingdom, 41–2, 46–7, 102, 170,
180–1, 226; cities, 182; migrants from,
237
United States, 3, 10, 43–4, 47, 98, 103,
105–6, 108–9, 114–15, 117–18, 145,
151–2, 169, 173, 177, 202, 216, 222, 226,
229, 233, 237, 240; Black population of
the, 153, 233, Black segment of the, 152;
children, 172–3; continental, 15;
long-term migrants to, 106; recent
migrants to, 106; South-west, 147, 229;
South-west, Clovis-point sites in the,
137; United States migrants to Hawaii,
233; urban areas of the, 110; Whites,
202, 233
urban environment (see also environment,
urban), patterns of biological adaptation
to the, 91
urban migrants (see migrants, urban)
urban migration (see migration, urban)
urban natives, 106, 111
urbanization, 91, 95–6, 106, 109, 115, 118,
217, 228; diseases of, 110; effects of, on
human health, 110; level of, 103;
'perversity of', 96, 118
urinary electrolytes, 113
Utah Mormons, 109

value: value system, 104; urban values, 118
Vannatu, 138
Virginia, 44
vital capacity, 11
vitamins: vitamin A, 187; vitamin D, 180,
182; deficiency of, 180; fortification of
food products, 181; provitamin D, 182

Wales, 42–4, 109
Wallace's: Deep, 18; Line, 16
water, 96; treatment, 93
weight, see body weight
West Africa, 58, 176, 220; Bantu, 153;
forced relocation of West African
populations to the New World, 177
West Indies, 178
Westernization, 168–9, 192, 194, 201–2,
217, 223; diseases of, 223; Westernized
children, 223
Whites in: New York, 115; South Africa,
226–7, 229–30; 233; the United
Kingdom, 226–7; United States, 202, 233

Wisconsin, 242
women, 106, 198, 230, 233, 241; fertility of
migrant, 105; fertility surveys of, 185;
Iranian, 225; migrant, 103, 105–7, 237;
Omaha and Seminole Amerindian, 105;
pregnant Pakistani, 182; rural- and
urban-living Zulu, 110; Seminole, in
urban areas, 106; urban migrant, 105;
urban Omaha, 105; women's education,
106
work, 190, 221; ability to perform, 11, 189;
capacity, 185, 189, 191, conditions, 223;
demands of, 11; environment, 220;
habits, 243

workers: agricultural, 111; blue-collar,
111; Hispanic migrant farmworkers, 242;
migrant, 220, 247; white-collar, 111;
women migrant, 242
World Health Organization (WHO),
238–9; standards, 112

Yemen, 216; Yemenites, 224, 226
Yukon: River, 17, 26–7, 30, 32; Territory,
26

Zaire, Katanga Province, Hutu children
in, 174–5
Zulus, blood pressure of, 110, 178, 225